THE
FUNDAMENTALS
OF
CRYSTALLOGRAPHY
&
MINERALOGY

THE FUNDAMENTALS OF CRYSTALLOGRAPHY & MINERALOGY

First Edition

SPEARS
MEDIA PRESS

DENVER

SPEARS MEDIA PRESS LLC

7830 W. Alameda Ave, Suite 103-247 Denver, CO 80226, USA

First Published in the United States of America by Spears Media Press LLC

www.spearsmedia.com

info@spearsmedia.com

Information on this title: www.spearsmedia.com/crystallography-mineralogy

© Aretas N. Ndimofor 2018

All rights reserved.

Ordering Information:
Special discounts are available on bulk purchases by corporations, associations, and others. For details, contact the publisher at any of the addresses above.

ISBN: 978-1-942876-24-3 [Paperback]
Also available in ebook (Kindle)

To all Geology/Earth Science Teachers and Students both present and future.

CONTENT

CRYSTALLOGRAPHY
Chapter 1

MINERALOGY
Chapter 2

Figures

Tables

PREFACE

In "**The Fundamentals of Crystallography and Mineralogy**", an attempt has been made to cover the essential requirements of the "O" and "A" level Geology Syllabi in Crystallography and Mineralogy. This is to keep abreast with the various aspects contained in the remobilized Concept of Competence Based Approach (CBA) put in place by the Ministry of Secondary Education (MINESEC) of Cameroon through the General Certificate of Education (GCE) Board, the Regional Delegations of Secondary Education and Teacher Associations.

Besides being suitable for secondary and high school students, the book also proves useful to first year students in Universities and colleges.

The text is enhanced by the literal use of some local materials for clarification, the use of diagrams in the form of graphs, photographs, tables etc. Hands-on activities and study questions to grasp Bloom's Taxonomy and the three learning domains are located at the end of each chapter.

On behalf of the author, I will like to thank all collaborators who supplied the useful inputs that made this book the masterpiece that it is.

Ngwa Cyprian Neba
Geology Teacher
GBHS BELO
10/04/2016

ACKNOWLEDGEMENT

In the preparing this book, I have greatly benefitted from the advice and encouragement of family members, relatives, colleagues and friends. Mentioning individual names will mean producing a separate book and giving it the title "**Acknowledgement**", so if you do not see your name here, do not feel less important.

I wish to thank the members and relatives of the following families: The Ndimofors, the Fons and the Taminangs. All the Geology teachers and students nationwide especially those of GHS Njinikom, GBHS Fundong, GBHS Belo, GHS Mbingo, GBHS Downtown Bamenda, St. Frederick Bamenda, GBHS Santa, CCAST Bambili and LBA Yaoundé. The entire North West Geology Teachers' Association (NOWEGETA) and its various zones especially the Boyo zone. The 42nd graduation Batch of ENS Annex Bambili especially those of the Diamond Family. The 2004 Upper Sixth Science students of GBHS Santa especially those of US4. Miss Nsom Vera Nsack, Mr. Ngole Nzalle and Muluh Sylvester (who assisted immensely in digitizing the manuscripts).

And finally the following very exceptional individuals, whose works have continued to inspire and motivate me; Mr. Forbeteh Moses (My very first geology teacher, presently North West Regional Pedagogic Inspector (RPI) for Geology, Mr. Duga Vincent, Mr. Tansida Henry and the TABS GROUP (who have done several geology related publications especially laboratory manuals), Mr. Formbui Emmanuel (Principal of GHS Njinikom), Mme Dorothy Suh Ndikum, Mme Rose Epie and the entire Monday Show Crew of CRTV.

This book is hard proof that all your prayers and support were not like water being poured on a duck's back, but were miracles that came as answers to my own prayers and energy to my spirit and muscles whenever I was tempted to give up. Although I have done my best to avoid any errors, they are inevitable in any human endeavour. I therefore encourage and welcome any useful corrections to this first edition.

By the Special Grace of God Almighty, I have reached this far and I am sure He is going to complete this good work He has begun in me, through Christ Jesus. Amen.

Aretas Ndimofor
Geology Teacher
GHS Njinikom

Aretas Ndimofor is an educator and a social entrepreneur from Akum, North West Region of Cameroon. He holds a BSc in Geology, Mining and Environmental Science from the University of Bamenda and a Teaching Diploma in Geology (Dipes I) from the Ecole Normale Supérieure (Annex), Bambili.

He is also a Microsoft Office Specialist Master (MOSM) in Office 2013 and 2016, a Microsoft Innovative Educator (MIE), a Google Certified Digital Marketer and Graduate from the School of Entrepreneurship and Economic Development Studies (SEEDS) of African Dreams (CEMMED). He has been teaching geology in secondary and high Schools since September 2009 and was the head of the Geology Department of Government High School Njinikom, Cameroon from 2013 to 2016. This textbook is inspired by his expertise in the area and his experiences teaching geology in schools in Cameroon.

He is founder of TC Techies (Twenty-first Century Techies), an initiative focused on the promotion of Digital Literacy and certification for youths, and CERTINED (Certified Innovative Educators), which trains educators on e-learning design.

He enjoys reading personal development books, listening to music and motivational tapes, dancing, creating e-learning tutorials, video and sound editing, networking and helping others in the area of personal development.

THE FUNDAMENTALS OF
CRYSTALLOGRAPHY & MINERALOGY

CHAPTER 1

Crystallography

Chapter Objectives

At the end of this chapter, you will be able to:

- Define some basic terms as used in geology such as; Crystallography, crystals, crystalline, amorphous, Euhedral, anhedral, unit cell, form, habit, face, edge, node, zone, zone axis, crystal symmetry, crystallographic axis, Miller indices, parameters, interfacial angles.
- State and define common crystal forms and habits.
- Explain the law of constancy of interfacial angles.
- Describe the crystallographic axes in the various crystal systems.
- Index crystal faces of common crystal forms.
- Outline the bases for the classification of crystals into systems and classes.
- Classify crystals into different systems and classes.

1.1 Introduction

Crystallography is the science that studies the geometry of crystals. That is the shape and forms that minerals assume in space in 3 dimensions (the regular patterns and interfacial angles of crystals).

1.1.1 Word History

The word crystal, was derived from a Greek word krustallos meaning 'clear ice' formed by the freezing of water. The ancient Greeks were amazed by Quartz (a mineral and rock) which occurred in forms having a characteristic shape and being bounded by flat surfaces (faces). From the transparency of Quartz, together with the presence of inclusions in it, it was thought that Quartz resulted from the freezing of water under intense cold and the name krustallos was given to it. The application of this name was later extended to all minerals that showed forms with smooth surfaces. These forms are crystals and their study is crystallography.

1.1.2 Definition

A crystal can be defined as a homogenous solid bounded by naturally formed plane (or smooth) surfaces called faces which can be related to a regular internal arrangement of atoms.

Crystals are formed by a process called crystallization during the solidification of minerals from the gaseous or liquid states or from solutions of magmatic origin

It is the regular internal arrangement of atoms within a mineral which really defines whether or not a mineral is crystalline. The study of the outward appearance of minerals (crystals) is important to geologists because it helps them identify and recognize different minerals. This is possible because different minerals have different chemical compositions and different internal atomic arrangements and therefore they have different crystal shapes and forms.

1.1.3 The Unit Cell

The structure of every crystal is a construction of atoms or groups of atoms arranged in three dimensional patterns which are repeated throughout the crystal.

A unit cell is the smallest complete unit of pattern of a crystal. Many unit cells combine in repetition to form a whole crystal for example;

Sodium Chloride (rock salt), a cubic mineral is an excellent example of a mineral with a cubic unit cell. Here, the unit cell consists of a cube with sodium ions at the corners and at the centre of the faces and chlorine ions at the centre of the edges of the cube as shown below:

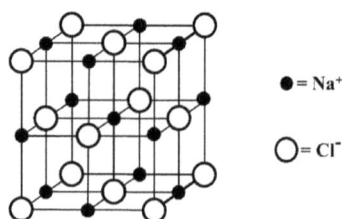

Figure 1: The Sodium Chloride Unit cell

The unit cell in sodium chloride can be interpreted as having 8 small cubes with either sodium or chlorine ions at the corners.

There are 7 major types of unit cells namely:
- The Cubic unit cell
- The Tetragonal unit cell
- The Hexagonal unit cell
- The Trigonal unit cell
- The Orthorhombic unit cell
- The Monoclinic unit cell
- The Triclinic unit cell

They give rise to seven systems under which crystals can be classified.

The external morphology of a crystal (like faces, edges, angles and form) and the general symmetry of any crystal are determined by the internal arrangement of these unit cells since they are the building block for crystals.

1.1.4 Crystallisation

Crystals can be formed by solidification from liquid or gaseous states or by precipitation from solutions saturated with ions. All these processes are called crystallization.

Crystallization is a process by which crystals are formed. Naturally occurring crystals (mineral crystals) are usually formed from a solution or a melt. In a liquid state or a solution, the atoms and ions are distributed haphazardly, but when the temperatures and pressure of the solution begin to drop, they quickly arrange themselves in an orderly manner forming crystalline solids called crystals.

A good example of crystallization can be seen during the formation of minerals from magma (molten matter). In the molten state, the magma contains ions which are randomly dispersed within the melt. As the temperature and pressure begin to drop, various ions are attracted to each other forming crystals of different minerals. If the process of cooling is slow and gradual, the ions will have enough time to migrate and come together thereby building large but few and well-formed crystals with smooth faces. However, if the drop in temperature is rapid, no time is allowed for the ions to move and coalesce (come together) so several centres or units of crystallization are developed around which many but irregularly oriented crystals are formed. Such crystals lack flat surfaces.

The degree of crystallization affects the development of crystals which is reflected on the external form and shape of the crystal.

Based on the degree of crystallinity and the development of external form of crystals, the following terms can be used to describe crystals:

i. *Euhedral*: It is a term that describes crystals that are well formed with smooth faces. Very few minerals show good forms. This indicates that they are formed under suitable conditions and such minerals are said to be crystallized.

ii. *Subhedral*: It is a term that describes crystals with partially formed or imperfectly formed faces. They are said to be crystalline.

iii. *Anhedral*: It is a term that that describes minerals that completely lack crystal faces.

According to the fineness of the grains of crystals, the following terms are used to describe minerals:

i. *Microcrystalline*: Describes minerals with fine grain aggregates, which can only be studied using microscopes.

ii. *Cryptocrystalline*: Describes minerals with crystal aggregates that are so fine grained that the individual grains cannot be seen using the microscope but can only be detected by the use of X-ray diffraction techniques. For example Chalcedony.

iii. *Amorphous*: These are minerals which lack any ordered internal arrangement of atoms for example Opal.

1.2 The External Features of A Crystal

The external features of a crystal comprise the following:

Faces, Edges, forms, solid angles (or nodes), habits and interfacial angles.

1.2.1 Faces

Faces are flat and smooth surfaces bounding or enclosing a crystal. The faces on a crystal are usually well arranged in a regular manner which is related to the internal arrangement of the atoms. There are two kinds of faces which are like and unlike faces

i. *Like faces*, are those which have the same size and shape in a regularly developed crystal (that is faces which have the same properties and dimensions).Like faces are common in minerals that crystallize in the cubic system.

ii. *Unlike faces* are those which do not have similar sizes and shapes. They therefore have different patterns of atoms within them.

1.2.2 Forms

Form refers to the assemblage of faces within a crystal which together may wholly or partially constitute the exterior of a crystal. Some crystals can be made up entirely of like faces while other crystals can be made up of like and unlike faces. The form of a crystal can be described as open or close.

i. *Closed form or Simple form:* A form is closed if it is made up entirely of like faces (faces which close space by themselves)for example a cubic form has 6 like faces.

ii. *Open form:* A form is said to be open if it has a combination of like and unlike faces for example Beryl has an open form made up of both pyramidal and prismatic faces.

1.2.3 Edges

An edge of a crystal is a line of intersection of 2 adjacent faces. The position of an edge depends on the position of the faces that produce the intersection.

1.2.4 Solid angles or Nodes

A solid angle or node is formed by the intersection of 3 or more faces.

1.2.5 Habit

This is the characteristic shape of a crystal produced by a variation in the number, size and shape of its faces. For example a cube would assume a cubic habit while octahedrons assume octahedral habits.

The habit of a crystal is influenced by the environment and conditions under which the crystal grew or was formed. For example Calcite usually has a rhombohedron habit but may assume a scalenohedron.

1.2.6 Interfacial angles

An *interfacial angle* is the angle between the normals or perpendiculars of 2 adjacent faces.

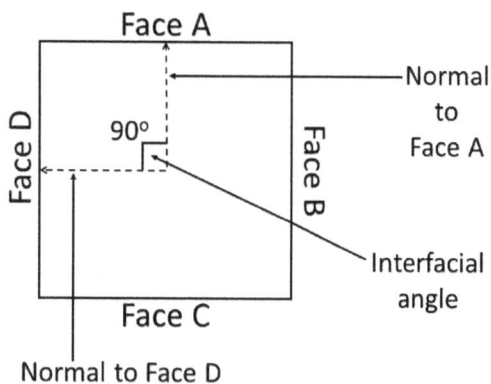

Figure 2: Interfacial Angle

Measuring Interfacial angles

Interfacial angles are measured using an instrument called a Goniometer. There are two kinds of goniometers which are the Contact and Reflective goniometers. The contact goniometer is used for large

crystals while the reflective goniometer is used for small crystals with flawless or smooth faces.

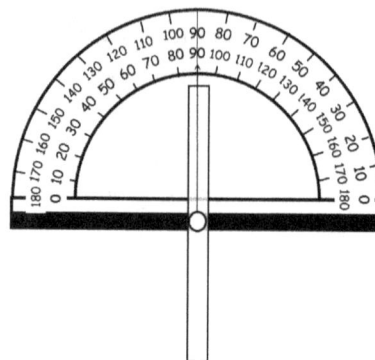

Figure 3: A Contact Goniometer

How the Contact Goniometer is used

The contact goniometer consists of two straight edged arms which are movable on a pivot or screw and connected to a graduated arc similar to that of a protractor.

The measurements are carried out by bringing the 2 arms firmly and accurately into contact with the adjacent faces of the crystal and the angle between them read off from the graduated arc.

The angle that is actually measured is the internal angle (perpendicular to the 2 faces) and must be subtracted from 180 degrees to obtain the interfacial angle.

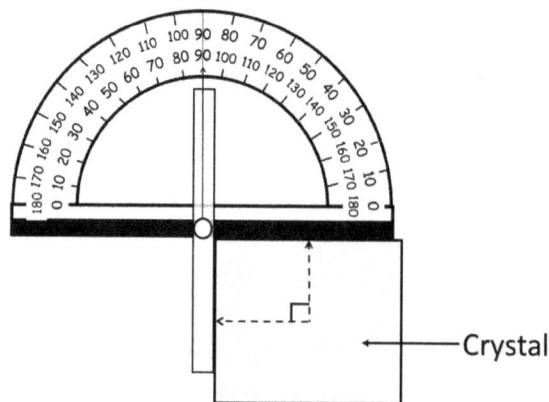

Figure 4: Using the Contact Goniometer

4

The Law of constancy of Interfacial angles

In 1669, Steno[1], after studying the interfacial angles on different samples of Quartz, put forward the law of constancy of interfacial angles which states that:

"For all crystals of a given mineral, the interfacial angles between any two given faces are always constant (the same)."

This law only holds if they are made up of the same chemical composition and measurements are done at the same temperature.

Geologists have been able to predict that crystals are built up by an orderly arrangement of atoms forming rows which correspond to crystal faces. The faces intersect at the edges which correspond to the intersection of the atomic rows within the crystals. This means that for all crystals of the same mineral, the faces and edges are always the same (as in the Quartz crystal). Careful studies of crystals have shown that the atomic structure for the crystals of any one mineral is fixed; consequently, the positions of the faces of such crystals are fixed. Thus based on measurements, the tendency for corresponding interfacial angles to be constant for all the crystals of a given mineral brings about *the law of constancy of interfacial angles*.

1.2.7 Zonation

A zone is a set of faces or a collection of faces which are parallel to a given direction. This direction is termed the zone axis.

A zone axis is a direction in which faces belonging to the same crystal zone are parallel to.

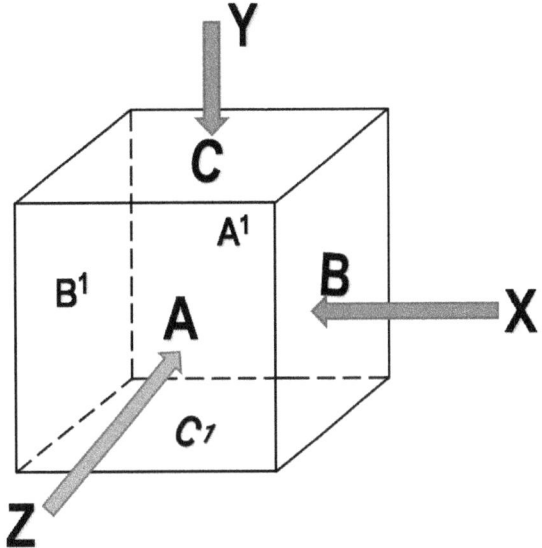

Figure 5: Zones and Zone axes

From above, there are three zone axes (X, Y and Z) and therefore the faces can be divided into three zones as follows:

- Zone 1 consists of faces A, C, A1, C1; the zone axis is X
- Zone 2 consists of faces A, B, A1, B1; the zone axis is Y
- Zone 3 consists of faces B, C, B1, C1; the zone axis is Z

Faces in the same zone must not necessarily be adjacent to each other. However, their edges must be parallel to each other.

1.3 The Elements of Symmetry

Crystals are built up of atoms which constitute the internal structure of the crystals. The unit cells within the crystal repeat themselves in a regular pattern causing most crystals to appear symmetrical.

Symmetry is defined as the degree of regular positions of common features (faces, edges, nodes, etc.) on a crystal. The degree of symmetry varies from one crystal to another and this variation has been used to classify crystals into systems and classes.

---

[1] Nicholas Steno was a Danish scientist (that is from Denmark)

The three important symmetry elements are:
- *Centre of symmetry,*
- *Axis of symmetry, and*
- *Plane of symmetry.*

1.3.1 Centre of symmetry:

It is an imaginary point within a crystal around which like features (faces, edges, nodes, etc.) on the crystal are arranged on opposite sides and are equidistant from it.

In crystals that have centres of symmetry, it is assumed that each feature on the crystal has a corresponding feature on the opposite side of an imaginary point in the middle called the centre of symmetry. Such features can be linked by a line passing through this point.

Most crystals especially those of the cubic system have centres of symmetry.

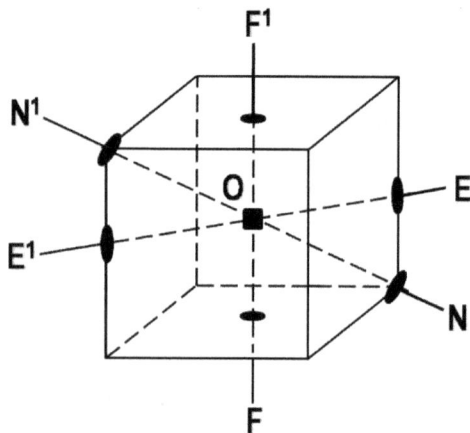

Figure 6: Demonstration of faces, edges and nodes on a cube

- F1-F indicates Faces,
- E1-E indicates Edges,
- N1-N indicates Nodes which are on opposite sides and Equidistant from the point O which is the centre of symmetry.

1.3.2 Plane of symmetry:

It is an imaginary line or position through which a crystal can be cut or divided into two equal and similarly placed halves such that one half is a mirror image of the other. In regularly formed crystals, the planes of symmetry are either axial or diagonal.

For example, a cube has 9 planes of symmetry (3 axial and 6 diagonal).

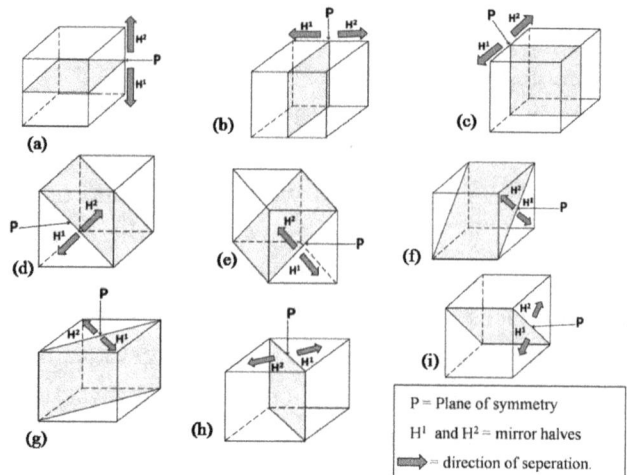

Figure 7: Planes of symmetry on a cube; a – c are axial planes while d – i are diagonal planes

1.3.3 Axis of symmetry:

It is an imaginary line passing through a crystal about which a crystal can be rotated through 360 degrees (a complete turn) so that the same features (faces, edges, or nodes) come to occupy the same position in space more than once.

The degree of regularity of a feature in space during one complete rotation is known as the fold. Fold of an axis of symmetry is the number of times any feature of a crystal occupies the same position during a complete rotation. That is 360 degrees/n (where n ≠ 1). n represents the degree of the axis.

There are 4 common axes of symmetry:

Fold	Other name	Symbol	n value
Two fold	Diad		$n = 2$ The same view occurs every 180° on rotation
Three fold	Triad		$n = 3$ The same view occurs every 120° on rotation
Four fold	Tetrad		$n = 4$ The same view occurs every 90° on rotation
Six fold	Hexad		$n = 6$ The same view occurs every 60° on rotation

Table 1.1: Common axes of symmetry and their symbols

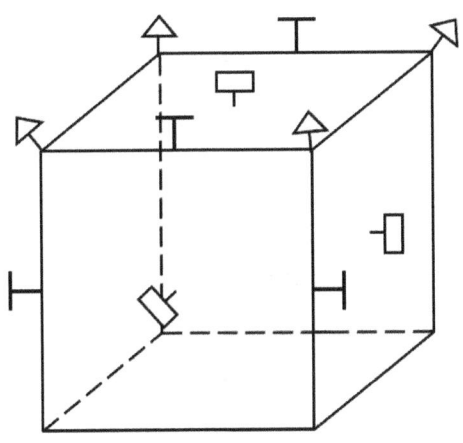

Figure 8: Axes of symmetry on a Cube

The symmetry elements are very important because they have been used to classify crystals into systems and systems into classes. All together there are 32 symmetry classes and 7 symmetry systems.

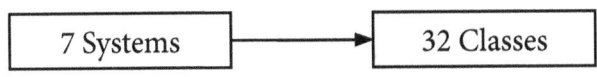

The class in a system that has the highest symmetry elements is known as the Holosymmetry class. For example; Galena falls under the Cubic system and has 32 symmetry elements.

Each system has a characteristic symmetry element which is a diagnostic property to that system while the other elements simply define the classes of that system.

The following table shows the different systems and their diagnostic symmetry elements:

S/N	System	Number of classes	Diagnostic symmetry element
1	The Cubic System	5	4^{iii}
2	The Tetragonal System	7	1^{iv}
3	The Hexagonal System	7	1^{vi}
4	The Trigonal System	5	1^{iii}
5	The Orthorhombic System	3	3^{ii}
6	The Monoclinic System	3	1^{ii}
7	The Triclinic System	2	None

Table 1.2: Diagnostic symmetry elements of the 7 crystal systems

- 50% of all known crystals belong to the monoclinic system
- 10% belong to the triclinic system
- 25% belong to the orthorhombic system.

Of the remaining 4 systems, the Cubic system is the most abundant followed by the Tetragonal, Trigonal and finally the Hexagonal system.

1.4 Crystallographic Axes and Crystallographic Angles

1.4.1 Crystallographic axes

These are directions in space chosen so that crystals belonging to the same system may be conveniently referred to by means of intercepts.

In other words, the axes are lines that run from a point of origin (O) which is the at the centre of crystal to the centre of faces and edges of the crystal which can be cut off (intersected) at various lengths to identify the crystal faces of the various systems. Most crystallographic axes are selected from amongst the axes of symmetry, therefore, all crystallographic axes are axes of symmetry but not all axes of symmetry are crystallographic axes.

Based on the angles at which the axes intersect each other, there are 3 main groups of axes as follows:

14.1.1 Orthogonal Axes:

These are crystallographic axes in which the three directions or axes lie mutually at right angles to each other. Examples of systems with this type of crystallographic axes are: the Cubic, Tetragonal and Orthorhombic systems.

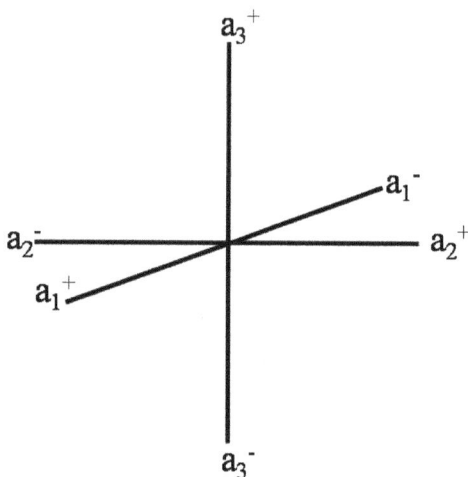

Figure 9: Orthogonal axes

1.4.1.2 Non-Orthogonal crystallographic axes

These are 3 directions in which one or more of the axes are not at right angles to each other. For example the Monoclinic and Triclinic systems.

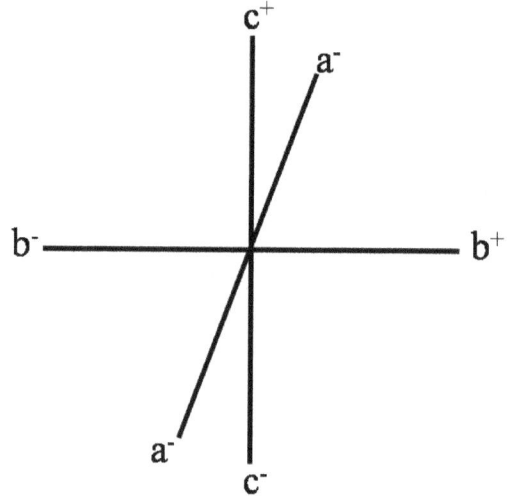

Figure 10: Non-Orthogonal axes

1.4.1.3 Special set

This constitutes 3 horizontal crystallographic axes or directions arranged mutually at 120 degrees to each other but at right angles to a fourth vertical axis. For example the Hexagonal and Trigonal systems.

<fig. 1.11 near here>

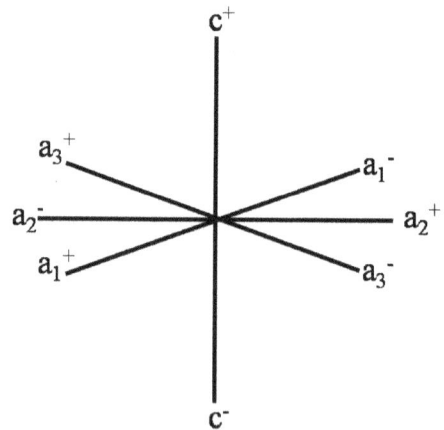

Figure 11: Special Set

Naming of Crystallographic Axes

There are certain signs and rules (conventions) used in the lettering and ordering of crystallographic axes. Generally the axes are labelled a, b and c on the crystal faces and edges of the various systems based on the relative distances from the origin within the unit form (the distance at which the crystal face intersects the crystallographic axis from the origin or centre of symmetry).

Where the unit form cuts (intersect) all crystallographic axes at equal lengths like in the Cubic system, the three crystallographic axes are interchangeable and are labelled a1, a2, a3 as demonstrated below:

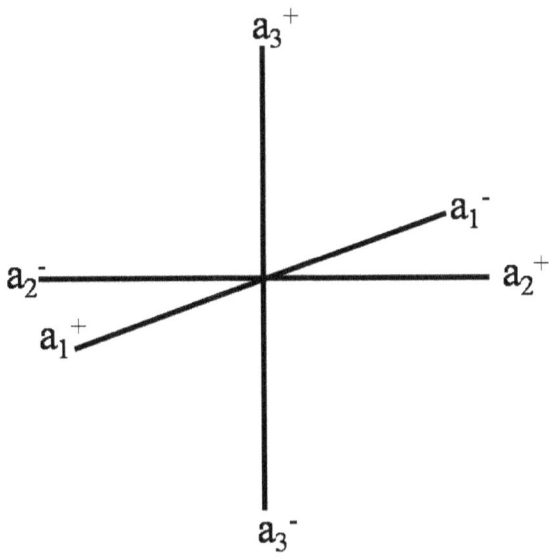

Figure 13: a1, a2, c crystallographic axes

Figure 12: a1, a2, a3 crystallographic axes

In some crystals, the unit form intersects the first 2 crystallographic axes at equal lengths and the third at a different length. In this case, the axes with the same lengths are labelled a_1, a_2 while the one with a different length is labelled c. Consequently, a_1 and a_2 are interchangeable. A typical example is shown by the Tetragonal system.

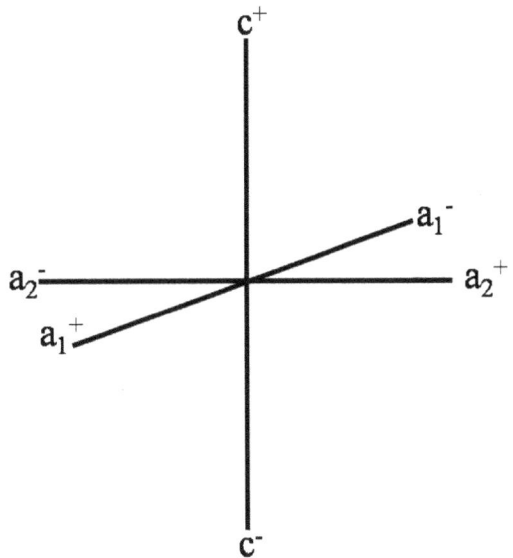

In a unit form where all the crystallographic axes are intersected at unequal lengths, the crystallographic axes are labelled a, b, c.

- The *a* crystallographic axis is horizontal and runs from front to back.
- The *b* crystallographic axis is also horizontal and runs from right to left.
- The *c* crystallographic axis is the vertical axis which runs from top (into the sky) to bottom (into the ground).

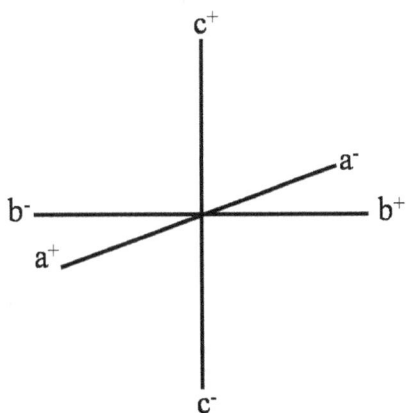

Figure 14: a, b, c crystallographic axes

Most crystals have positions in space that can be referred to by using only 3 crystallographic axes, but there are some crystals that can be referred to by using 4 crystallographic axes denoted a_1, a_2, a_3 and c. The first 3 crystallographic axes (a_1, a_2 and a_3) have equal lengths are interchangeable and are horizontal while the fourth crystallographic axis (c) has a different length and is vertical. This case is found in the Hexagonal system.

The "a" crystallographic axes are given positive (+) and negative (-) signs to indicate their directions from the origin (centre of the unit cell). One end of each axis is positive (+) and the other is negative (-) on opposite sides of the origin as follows:

- The "a" crystallographic axis is considered positive (a^+) forward from the origin and negative (a^-) backward.
- The "b" crystallographic axis is considered positive (b^+) to the right of the origin and negative (b^-) to the left.
- The "c" crystallographic axis is considered positive (c^+) upward (to the sky) from the origin and negative (c^-) downwards (into the ground) from the origin.

1.4.2 Crystallographic angles

These are the angles formed between the crystallographic axes at the point of intersection (the origin). They are denoted alpha (α), beta (β) and gamma (γ)

- The angle between the *b* and *c* crystallographic axes is alpha (α).
- The angle between the "*a*" and *c* crystallographic axes is beta (β).
- The angle between the "*a*" and *b* crystallographic axes is gamma (γ).

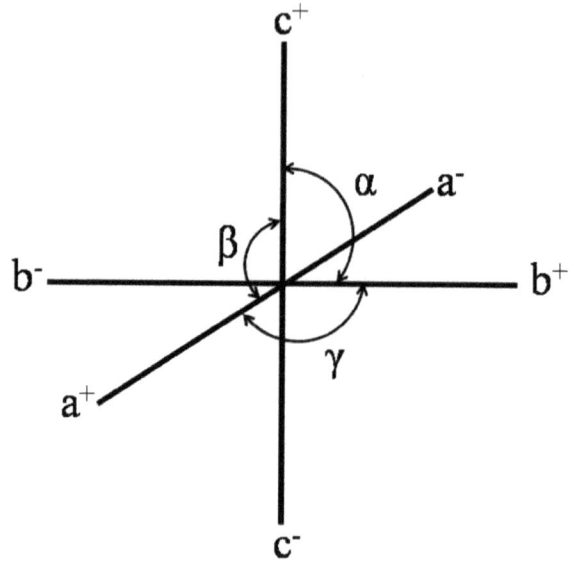

Figure 15: The a–b, a–c and b–c crystallographic angles

1.5 Crystallographic Notation (Intercepts, Parameters and Indices)

1.5.1 Intercepts

In describing crystal faces, it is important to indicate the crystallographic axis is intersected or not intersected by the faces.

Intercepts are where crystallographic axes are met and intersected by crystal faces.

The point of intersection[2] of the crystallographic axis is called the origin and is given the value 0. Reference is usually taken from that point. Based on this idea, intercepts refer to the distance of a crystallographic axis from the origin to a crystal face, within a unit cell. Usually, the intercepts of a crystallographic axis in a unit form is always 1 but there are a few variations.

For example in the figure below, the face (labelled 1) intersects the a crystallographic axis at a unit length (OS) and is parallel to the b and c crystallographic axes. The intercepts of this face will be $1a$, ∞b, ∞c.

---

[2] To intersect means to cross something, or to cross each other.

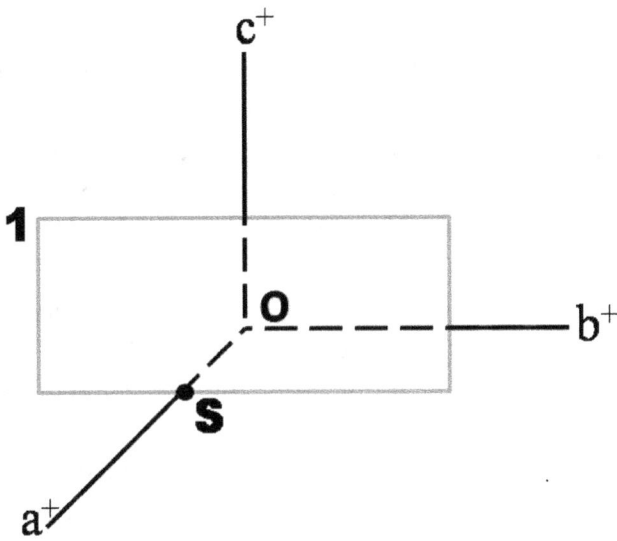

Figure 16: Face 1, intercepting the a crystallographic axis and parallel to the other two, (1a, ∞b, ∞c)

The next figure shows two faces; 2 and 3 which both intersect the "*a*" crystallographic axis and are parallel to both the *b* and *c* crystallographic axes. Face 3 intersects the "a" crystallographic axis at a distance, OS¹, which is further than the distance at which face 1 intersects the same axis. Its distance is considered to be twice the unit length and its intercepts will therefore be 2*a*, ∞*b*, ∞*c*, relative to face 1 and the intercepts of face 1 will be 1*a*, ∞*b*, ∞*c*.

which intersect the same crystallographic axis at different lengths.

The next figure demonstrates a face, 4, which intersects both the "a" and "b" crystallographic axes at a unit length (OP and OP1 respectively) and is parallel to the c crystallographic axis.

Its intercept value can be given as 1a, 1b, ∞c.

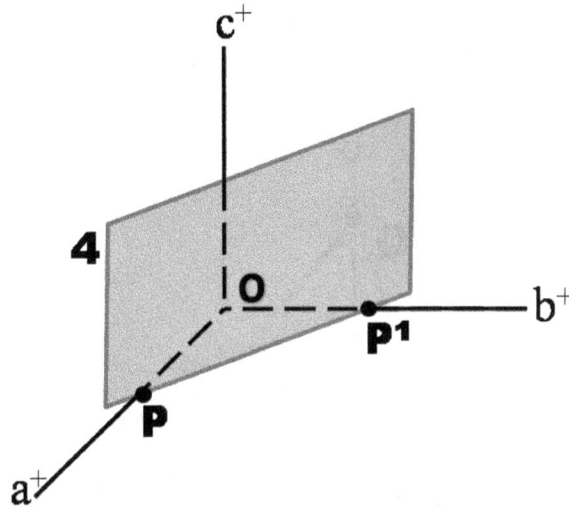

Figure 18: Face 4 intersects both the a and b crystallographic axes at a unit length and is parallel to the c crystallographic axis.

In some cases, a face can intersect all three crystallographic axes at unit length, in below, face 5 intercepts all crystallographic axes at unity (OK, OL and OM) giving it the intercepts 1a, 1b, 1c.

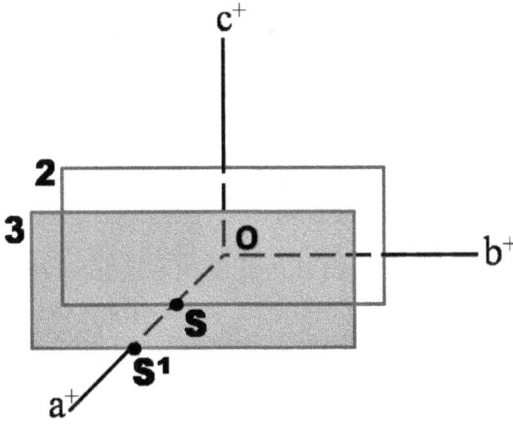

Figure 17: A comparison between 2 faces, 2 and 3

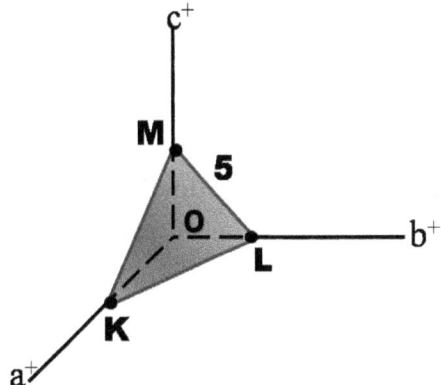

Figure 19: Face 5 showing intercepts 1a, 1b, 1c.

The next demonstration shows face 6 also intersecting all crystallographic axes but at unequal lengths compared to face 5. It intersects the "a" and c crystallographic axes at further lengths and the b crystallographic axis at the same length as face 5.

For face 5, the intercepts are 1a, 1b, 1c.

For face 6, the intercepts are 2a, 1b, 2c. The most widely used systems of notation are the Parameter system of Weiss[3] and the Index System of Miller[4].

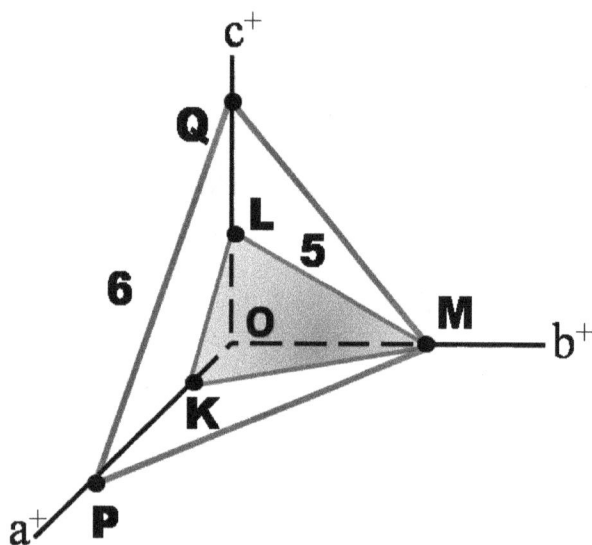

Figure 20: A comparison between 2 faces, 5 and 6 which intersect all crystallographic axes at different lengths.

1.5.2 Parameters (The Parameter system of Weiss)

Parameters are defined as the ratios of distances from the origin at which crystal faces cut crystallographic axes (for short, Parameters are ratios of intercepts).

Weiss considered the crystallographic axis to be labelled as *a*, *b*, *c*, for unequal axes, *a*, *a*, *c* for 2 equal

and 1 unequal axis and a, a, a, for all 3 axes being equal in length.

The intercept values that a face makes with the crystallographic axes are written before the axes.

For example: n*a*, m*b*, p*c*, where n, m, p are the parameter values when compared to the corresponding lengths cut off by the unit form (I I I).

NB: Unit form is the form whose face intersects the crystallographic axes at unit lengths (I I I) where reference is taken for the intercepts of the other crystal forms on the same axes.

In order to represent infinity or a face that is parallel to an axis (does not intersect an axis), Weiss used the sign ∞.

Thus following the Weiss parameters;

i.   *a*, 2*b*, ∞*c*, means that the face:
  • Cuts the a crystallographic axis at a distance of 1 unit (the same unit as the unit form),
  • Cuts the *b* crystallographic axis at twice the unit of the unit form and is parallel to the *c* crystallographic axis.

ii. *a*, ∞*b*, ∞*c*, means that the face:
  • Cuts the a crystallographic axis at a distance of 1 unit (the same unit as the unit form) and is parallel to the *b* and *c* crystallographic axes.

1.5.3 Indices (The Index System of Miller or Miller Indices)

Miller indices are a series of whole numbers used to indicate the precise positions or relative distance at which crystal faces intersect different crystallographic axes in space.

In this system of notation, the indices or reciprocals of the parameters are used. They are written in the order of crystallographic axes (that is the first number stands for the a crystallographic axis, the second number stands for the *b* crystallographic axis and the third number stands for the a crystallographic

---

[3] Christian Samuel Weiss (Feb. 26, 1780 – October 1, 1856) was a German Mineralogist

[4] William Hallowes Miller (Apr. 6, 1801 – May 20, 1880)

axis)

NB: In crystal systems that have 4 crystallographic axes, the first three numbers stand for the horizontal axes; a_1, a_2, a_3 and the fourth number stands for the vertical axis which is the c axis.

The whole numbers are taken from parameters and are always given in their most simple form by inversion and clearing off fractions.

For example, consider a face with Weiss parameters a, $2b$, ∞c

- The reciprocals of the parameters will be $\frac{1}{1}$, $\frac{1}{2}$, $\frac{1}{\infty}$ or $\frac{1}{1}a$, $\frac{1}{2}b$, $\frac{1}{\infty}c$
- By inversion, we will have $\frac{1}{1}$, $\frac{2}{1}$, $\frac{\infty}{1}$
- Simplifying it gives the Miller indices as 1 2 0, read as one, two, naught (zero) and *NOT one hundred and twenty.*

Similarly, a face which intersects the "*a*" crystallographic axis and is parallel to the *b* and *c* crystallographic axes, notated in Weiss parameters as a, ∞b, ∞c, will have the Miller index 100, read as one, zero, zero.

- A Weiss Parameter of $\frac{1}{2}$, 1, $\frac{1}{3}$, simplified into Miller Indices will be (2 1 3); Two, one, three.

Miller indices are very important because they show the relationship between a crystal face and a crystallographic axis by giving the relative length of each crystallographic axis from the origin (0) to the face in question. The larger the index, the smaller the corresponding intercepts and the closer the face is to the origin. On the other hand, the smaller the value, the farther the face is from the origin.

Take for instance (2 1 3) means that the face in question intersects:

- The *a* crystallographic axis at ½ the unit distance,
- The *b* crystallographic axis at a unit distance and
- The *c* crystallographic axis at ⅓ the unit length.

Since crystallographic studies at this level do not deal with measurements, most faces that cut all three crystallographic axes are considered to be at unity (I I I).

Any face which cuts the negative end of a crystallographic axis is indicated by putting a bar (-) above the index of that axis. For example (I Ī I), read as one, bar one, one, indicates that the face intersects

- The *a* crystallographic axis at the positive end,
- The *b* crystallographic axis at the negative end and
- The *c* crystallographic axis at the positive end.

The notation of crystal form is a summary of the crystal faces and written in braces, that is { } to differentiate them from Miller indices which are written in parenthesis, that is ( ).

NB: The number of faces in a Form depends on the symmetry of the crystal (that is the level of regular repetition of the same unit cell).

For example:

(I I I) represents a crystal face that cuts all three crystallographic axes (*a, b, c*) at their positive ends (known as a Pyramidal face),

While {I I I} represents a crystal form with 8 faces with each face cutting all three crystallographic axis (known as an Octahedron).

Notations of crystal forms are called *Form symbols.*

Form symbols have no bars on them. For example {I Ī I} is WRONG!

1.6 Crystal Forms

A crystal form refers to a set of crystal faces which have the same symmetry elements (or crystal faces that are related by symmetry).

Simple crystals may have only one crystal form but more complex and complicated crystals may have a combination of several different forms.

1.6.1 Some common forms in the Non – Isometric System

i. Pinacoids (or Parallelohedrons)

It consists of 2 parallel faces which intersects one crystallographic axis and are parallel to the others

- { I 0 0 } is the form symbol for a Front Pinacoid
- { 0 I 0 } is the form symbol for a Silica
- { 0 0 I } is the form symbol for a basal Pinacoid

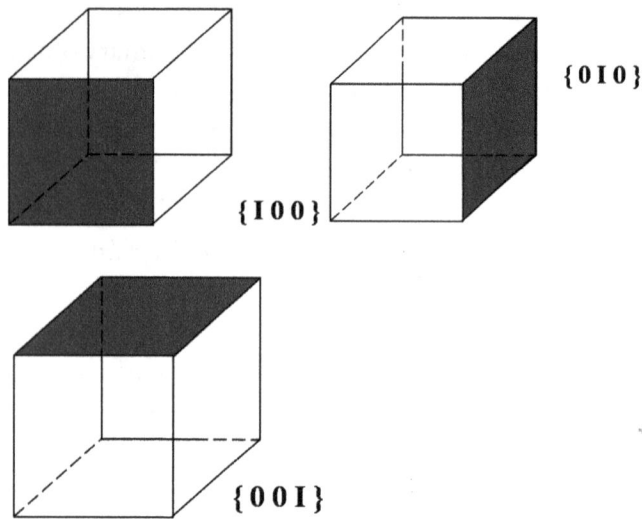

Figure 21: Pinacoids (front, side and basal)

ii. **Prisms**

It is a crystal form composed of 3, 4, 6, 8 or 12 faces, all of which are parallel to the c crystallographic axis and intersect one or two other horizontal axes. Examples of prisms include Trigonal Prisms, Tetragonal prisms, Rhombic prisms and hexagonal prisms. Some prisms are called *Prisms of the 1st Order* while others are called *Prisms of the 2nd Order*.

- Prisms of the 1st order are those whose faces intersect two horizontal crystallographic axes and are parallel to the rest. This is because the crystallographic axes are intersected by edges (the point where to faces meet).
- Prisms of the 2nd order are those whose faces

intersect one horizontal crystallographic axis and are parallel to the rest. This is because the crystallographic axes are intersected by faces.

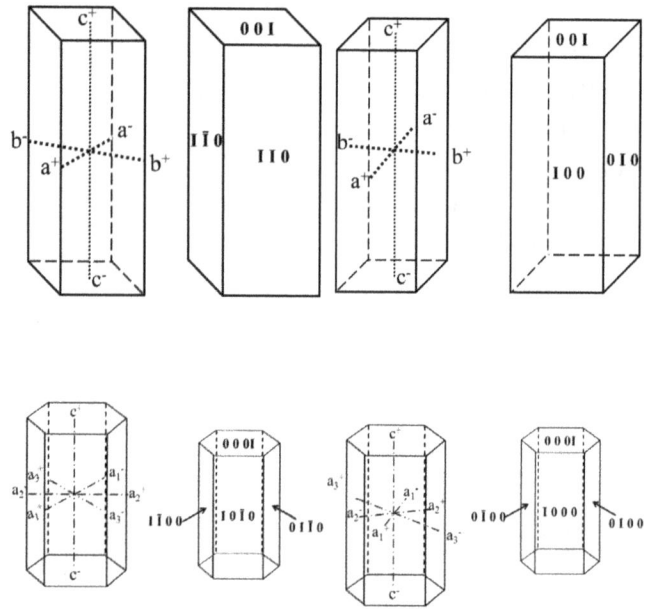

Figure 22: Tetragonal and Hexagonal Prisms (showing the difference between 1st and 2nd order prisms).

iii. **Dome (or Dihedron)**

It is an open crystal form consisting of 2 non – parallel faces. Each face cuts the c crystallographic axis and one other horizontal crystallographic axis and is parallel to the other. {I0I} or {0II}.

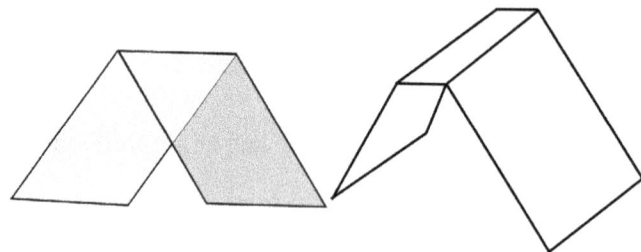

Figure 23: A dome

iv. Pyramid

It is an open form composed of 3, 4, 6, 8 or 12 non – parallel faces that meet at a common point. Each face intersects all three crystallographic axes. The Tetragonal, Trigonal, Orthorhombic, and Hexagonal systems all produce pyramids.

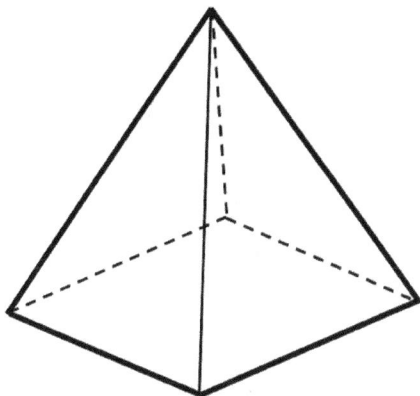

Figure 24: A Pyramid

v. Bipyramid:

It is an open form with 6, 8, 12, 16 or 24 triangular faces that intersect all 3 crystallographic axes. Just like the pyramid, there are also Tetragonal, Trigonal, Orthorhombic, and Hexagonal bipyramids.

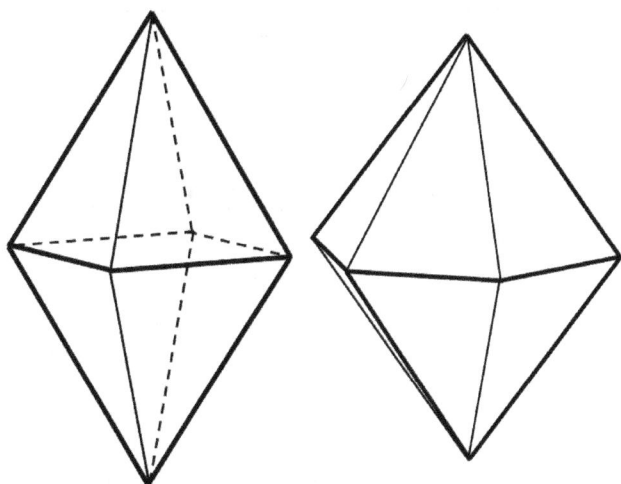

Figure 25: Bipyramids

vi. Scalenohedrons:

It is a closed form with 8 or 12 faces grouped in symmetrical pairs. There are Tetragonal Scalenohedrons (with 8 faces) and Hexagonal scalenohedrons with (12 faces). Scalenohedrons are built up of scalene triangles.

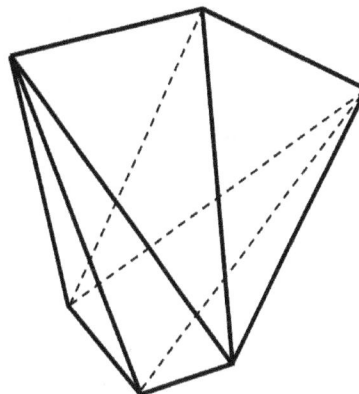

Figure 26: A scalenohedron

vii. Rhombohedrons

It is a closed form composed of 6 rhomb shaped faces in which each face intersects 3 crystallographic axes and is parallel to the fourth.

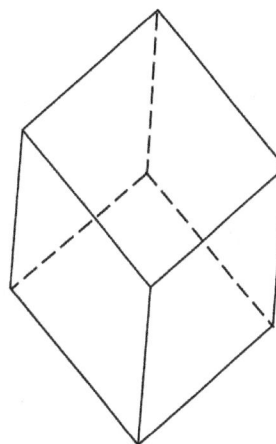

Figure 27: A Rhombohedron

1.6.2 Some common forms in the Isometric system
i.   Cube

It is a form with 6 square faces at 90° angles to each other. Each face intersects 1 crystallographic axis and is parallel to the other two. Its form notation is {00I}

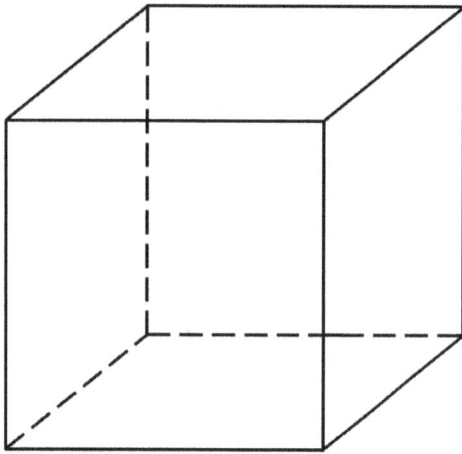

Figure 28: A Cube

Octahedron

It is a form composed of 8 equilateral triangles. These triangle – shaped faces intersect all 3 crystallographic axes at equal lengths (the same distance from the origin).

Its form notation is {III}.

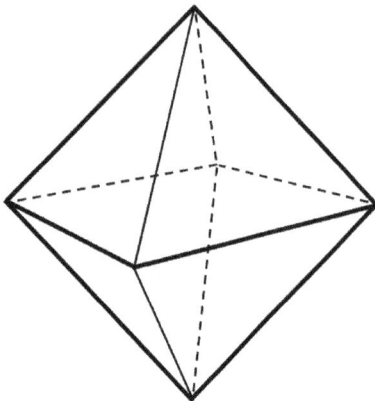

Figure 29: An Octahedron

ii.   Dodecahedron (Rhombic Dodecahedron):

It is a form composed of 12 rhomb – shaped faces each of which intersects 2 crystallographic axes and is parallel to the 3rd axis. Its form notation is {0 I I}.

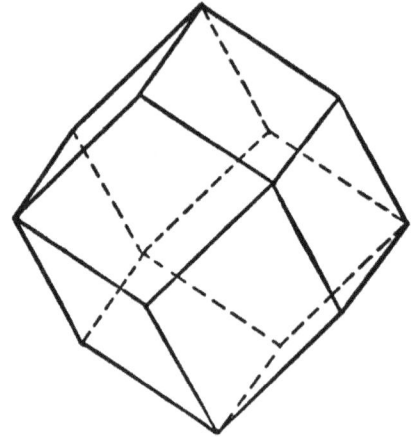

Figure 30: Dodecahedron (Rhombic Dodecahedron)

1.7 Classification of Crystals

As earlier mentioned (under elements of symmetry), the symmetry elements are very important because they have been used to classify crystals into systems and systems into classes. All together there are 32 classes under 7 symmetry systems.

7 Systems	→	32 Classes

1.7.1 Classification into systems:

The classification of crystals into systems is done on the basis of the following sub categories:
- The crystallographic axes,
- The crystallographic angles,
- The type of unit cell (unit form),
- The diagnostic symmetry element.

a). The crystallographic axes:
- In this sub category, note is taken on
- The number of axes (whether there are 3 or 4 axes) and

- The lengths at which the unit form intersects each crystallographic axis,

b). The crystallographic angles,

Here, the angles between the crystallographic axes are considered (that is whether they are orthogonal or non–orthogonal axes).

c). The type of unit cell (unit form):

In this sub category, the type of unit form that makes up the crystal is considered.

d). The diagnostic symmetry element:

Under this, the symmetry element which is common to all the classes under the system is considered. For example, the cubic system has 4^{iii} folds in all its classes, while the tetragonal system has 1^{iv} folds in all its classes.

1.7.2 Classification into classes:

The division of crystal systems into classes is based on the symmetry elements (plane of symmetry, axis of symmetry and centre of symmetry).

Although there are seven (7) crystal systems which are sub-divided into 32 classes, many of these classes have no mineral representative, or are represented either by very rare minerals or by chemical compounds. Most common minerals fall under one of the first 15 classes.

The *Holohedral* or *Holosymmetric* or *Holomorphic* class is the class in a system which contains the highest number of symmetry elements. Such a class can be used to represent that system. For example the Galena class in the Cubic system has 23 symmetry elements.

1. The Cubic System:

- It has 3 crystallographic axes all of equal lengths that is a = b = c, labelled ; a1 a2 a3
- Its axial angles are such that $\alpha = \beta = \gamma = 90°$

- The unit form is a solid with six square faces (a cube)
- The diagnostic feature for the system is 4iii fold.

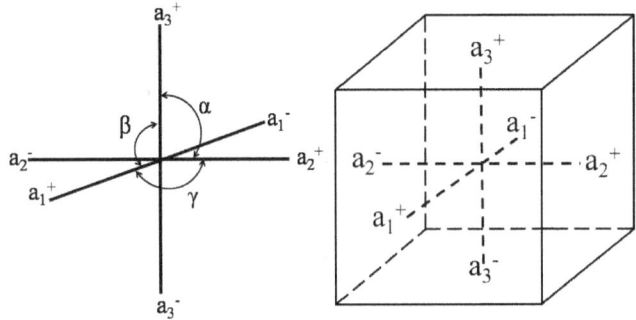

Figure 29: The crystallographic axes of the cubic system

The Cubic system has 15 classes with the holomorphic class being the Hexoctahedral class.

i. The Cubic Holosymmetric or (Hexoctahedral) class or Galena type:

It is the holosymmetric class of the Cubic system possessing 23 elements;
- A centre of symmetry
- 13 axes of symmetry (4^{iii}, 3^{iv} and 6^{ii})
- 9 planes of symmetry (3 axial and 6 diagonal)
- Examples of minerals under this class include: free metals such as Gold, Silver, Copper, Lead, Platinum, Iron, Halite (NaCl), Galena (PbS), Fluorite (CaF_2), and Spinels including Magnetite (FeO_4)

a. Didodecahedral Class or Diploidal class or Pyrite type:
- A centre of symmetry
- 3 planes of symmetry
- 7 axes of symmetry (4^{iii}, 3^{ii})
- Examples of minerals under this class include;

Pyrite (FeS2) and many Nitrates.

b. **Hexatetrahedral class or Tetrahedrite type:**
- No centre of symmetry
- 6 planes of symmetry
- 7 axes of symmetry (4^{iii} and 3^{ii}).
- Examples of minerals under this class include; Sphalerite or zinc blende (ZnS) and Phosphate.

c. **Pentagonal or icositetrahedral class:**
- No centre of symmetry
- No plane of symmetry
- 3 axes of symmetry (4^{iii}, 3^{iv} and 6^{ii})
- Tetrahedral Pentagonal Dodecahedral class:
- No centre of symmetry
- No plane of symmetry
- 7 axes of symmetry (4iii and 3iv)

Forms in the Isometric System

1. **Cube:** It is a form with 6 square faces at 90 degree angles to each other. Each face intersects 1 crystallographic axis and is parallel to the other two. Its form notation is {00I}
2. **Octahedron:** It is a form composed of 8 equilateral triangles. These triangle – shaped faces intersect all 3 crystallographic axes at equal lengths (the same distance from the origin). Its form notation is {III}.
3. **Dodecahedron (Rhombic Dodecahedron):** It is a form composed of 12 rhomb – shaped faces each of which intersects 2 crystallographic axes and is parallel to the 3rd axis. Its form notation is {0II}.
4. **Tetrahexahedron:** It comprises 24 isosceles triangular faces each of which intersects 2 crystallographic axes; 1 at unity, the other at a different length and is parallel to the third.
5. **Tetrahedron:** it comprises 4 equilateral triangular faces, each of which intersects all 3

crystallographic axes at equal lengths.

2. ***The Tetragonal System***
- The crystals in this system are referred to by 3 crystallographic axes which are perpendicular (90 degree) to each other.
- The two horizontal axes are equal in length and labelled a_1, a_2 while the third axis is either longer or shorter and labelled c. That is $a = b \neq c$ ($a_1 = a_2 \neq c$)
- Its axial angles are such that $\alpha = \beta = \gamma = 90°$
- The unit form is a system made up of straight prisms with square bases.
- The diagnostic property if this system is a unique 1^{iv} axis on the c crystallographic axis.

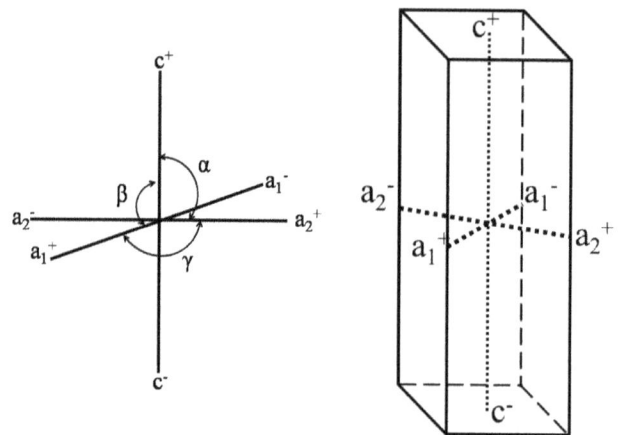

Figure 30: The crystallographic axes of the tetragonal system

The Tetragonal system has 7 classes as follows:
a. The Tetragonal holosymmetric (Ditetragonal bipyramidal) class or the Zircon type.
b. The Tetragonal hemimorphic (Tetragonal pyramidal) class
c. The Tetragonal Sphenoidal Class
d. The Tetragonal Bipyramidal class
e. The Ditetragonal hemimorphic (Ditetragonal pyramidal) class
f. The Tetragonal scalenohedral (tetragonal

bisphenoidal) class
g. The Tetragonal Trapezohedral class.

The holohedral class of the system is the *Tetragonal holosymmetric (Ditetragonal bipyramidal) class or Zircon type* with the following elements of symmetry:
- A centre of symmetry
- 5 planes of symmetry (3 axial and 2 diagonal)
- 5 axes of symmetry (1^{iv} and 4^{ii})

Forms in the Tetragonal System
1. **Basal pinacoid:** It is a form composed of 2 parallel faces perpendicular to the c crystallographic axis. The basal pinacoids exist in combination with the various prisms of this system which include:
    - Tetragonal Prisms of the 1st order {I I 0} and
    - Tetragonal Prisms of the 2nd order {0 I 0 }
    - **NB: Prisms of the 1st order are those whose faces intersect one horizontal axis and are parallel to the rest while Prisms of the 2nd order are those whose faces intersect two horizontal axes and are parallel to the rest.**
2. **Bipyramids {II0}**, which is a combination of prisms, bipyramids and pinacoids;

Tetragonal bipyramids of 1st order whose faces intersect all 3 crystallographic axes symbolized as {III}

Tetragonal bipyramids of 2nd order whose faces intersect the vertical crystallographic axis and 1 other horizontal crystallographic axis and is parallel to the other horizontal axis.

Examples of some minerals with this form include: *Idocrase, Zircon, Rutile and Cassiterite.* Of them all, Idocrase shows a combination of all the forms in the tetragonal system.

3. The Hexagonal System
- The crystals of this system are referred to by 4 crystallographic axes;
- 3 horizontal axes of equal lengths at 120 degrees to each other labelled a_1 a_2 a_3 and a fourth vertical axis labelled c which is perpendicular to the horizontal axes.
- **NB: the a_3 crystallographic axis is positive behind the origin and negative in front of the origin.**
- Its axial angles are such that $\alpha = \beta = 90$ degrees and $\gamma = 120$ degrees.
- The unit structure is a prism with a hexagonal base.
- The diagnostic property of this system is a unique 1^{vi} axis on the *c* crystallographic axis.

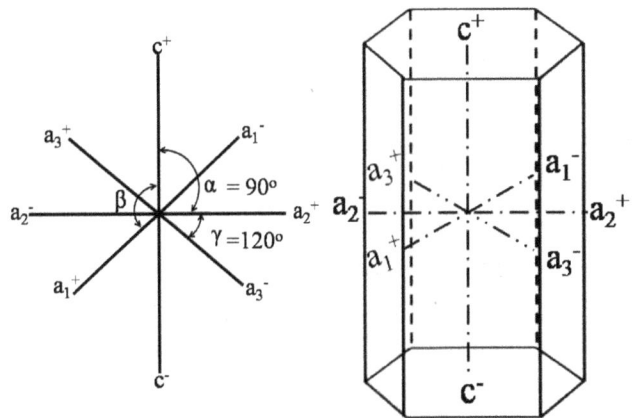

Figure 31: The crystallographic axes of the hexagonal system

The Hexagonal system has 7 classes as follows:
a. The Hexagonal holosymmetric (Dihexagonal bipyramidal) class or the Beryl type.
b. The Hexagonal hemimorphic (Hexagonal pyramidal) class
c. The Trigonal Bipyramidal class
d. The Hexagonal Bipyramidal class
e. The Dihexagonal hemimorphic (Dihexagonal pyramidal) class

f. The Ditrigonal Bipyramidal class
g. The Hexagonal Trapezohedral class.

The holohedral class of the system is the *Hexagonal holosymmetric (Dihexagonal bipyramidal)* class or the Beryl type with the following elements of symmetry:

- A centre of symmetry
- 7 planes of symmetry; 4 axial and 3 diagonal
- 7 axes of symmetry; 1^{vi} and 6^{ii}
- An example is Beryl ($BeAl_2Si_6O_{18}$)

Forms in the Hexagonal system
1. Basal pinacoids, $\{0\,0\,0\,I\}$
2. Hexagonal Prisms of 1st and 2nd order, $\{I\,0\,I\,0\}$ and $\{I\,I\,2\,0\}$
3. A dihexagonal Prism made up of 12 faces.
4. Hexagonal bipyramids of 1st and 2nd order.

4. The Trigonal System
- The axes of the Trigonal system are identical to those of the Hexagonal system. That is:
- 4 crystallographic axes;
- 3 horizontal axes of equal lengths at 120 degrees to each other labelled $a_1\,a_2\,a_3$ and a fourth vertical axis labelled c which is perpendicular to the horizontal axes.
- Its axial angles are such that $\alpha = \beta = 90$ degrees and $\gamma = $ *120 degrees.*
- The basic unit form of crystals in this system is a Rhombohedron.
- The systems diagnostic property is a unique 1^{iii} fold axis.

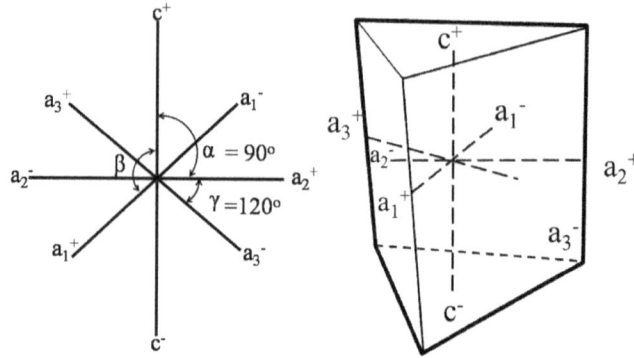

Figure 32: The crystallographic axes of the Trigonal system

The Trigonal system has 5 classes as follows:
a. The Trigonal holosymmetric (Ditrigonal scalenohedral) class or calcite type.
b. The Trigonal hemimorphic (Trigonal pyramidal) class
c. The Rhombohedral class
d. The Ditrigonal hemimorphic (Ditrigonal pyramidal) class or the tourmaline type.
e. The Trigonal Trapezohedral class or the Quartz type.

The holohedral class of the system is the Trigonal holosymmetric (Ditrigonal scalenohedral) class with the following elements of symmetry:
- A centre of symmetry
- 3 planes of symmetry
- 4 axes of symmetry; 1^{iii} and 3^{ii}
- For example; *Carbonates such as Calcite and Siderite, Hematite (Fe_2O_3) and Brucite (Mg(OH)$_2$).*

5. The Orthorhombic System
- This system has 3 perpendicular crystallographic axes with unequal lengths. That is
- $a \neq b \neq c$.
- Its axial angles are such that $\alpha = \beta = \gamma = 90°$
- The unit form for this system is a straight

prism with a rectangular base.
- The diagnostic property of the orthorhombic system is 3^{ii} along the c crystallographic axis.

The orthorhombic system has 3 classes as follows:
a. The orthorhombic holosymmetric (orthorhombic bipyramidal) class
b. The orthorhombic hemimorphic (orthorhombic pyramidal) class
c. The orthorhombic sphenoidal class.

The holohedral class of the system is the orthorhombic holosymmetric (orthorhombic bipyramidal) class with the following elements of symmetry:
- A centre of symmetry
- 3 planes of symmetry
- 3 axes of symmetry; 3^{ii}.
- Examples include: Barite, Amphiboles, Andalucite, Sillimanite, Olivine group.

Forms in the Orthorhombic system
- Pinacoids
- Prisms
- Bipyramids

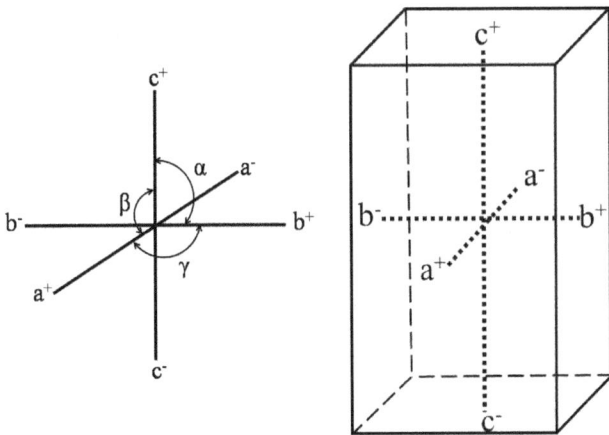

Figure 33: The crystallographic axes of the orthogonal system

6. The Monoclinic System

The Monoclinic system has the following characteristics;
- 3 crystallographic axes with unequal lengths intersecting at oblique angles.

That is $a \neq b \neq c$.
- The "a" crystallographic axis is inclined towards the "c" crystallographic axis. That is
- Its axial angles are such that $\alpha = \gamma \neq 90° \neq \beta$
- The unit form in this system is a solid with a rectangular prism and a rectangular base.
- The diagnostic property of the monoclinic system is 1^{ii}.

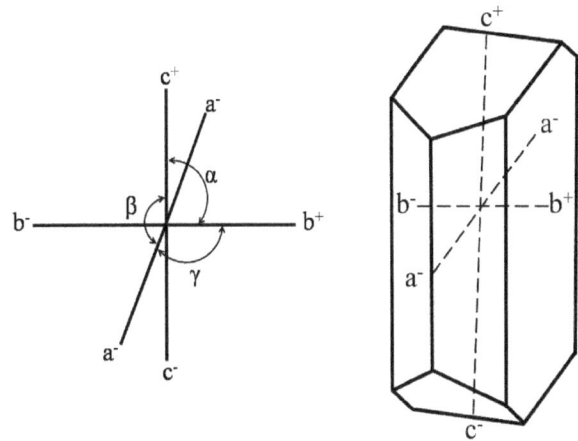

Figure 34: The crystallographic axes of the monoclinic system

The monoclinic system has 3 classes as follows:
a. The monoclinic holosymmetric (monoclinic prismatic) class
b. The monoclinic hemimorphic (monoclinic sphenoidal) class
c. The monoclinic clinohedral (monoclinic domatic) class

The holosymmetric class of the system is the monoclinic holosymmetric (monoclinic prismatic) class with the following properties:
- A centre of symmetry

- 1 plane of symmetry
- 1 axis of symmetry; 1^{ii} along the b crystallographic axis.
- Examples include: *Gypsum, Orthoclase, Augite, Hornblende, Micas, Epidote group and Chlorite group.*

7. The Triclinic System
This system has the following characteristics:
- 3 unequal crystallographic axes all at oblique angles.

That is;
- $a \neq b \neq c$.
- Its axial angles are such that $\alpha \neq \gamma \neq \beta \neq 90°$
- The diagnostic property of this system is the lack of axes and planes of symmetry.

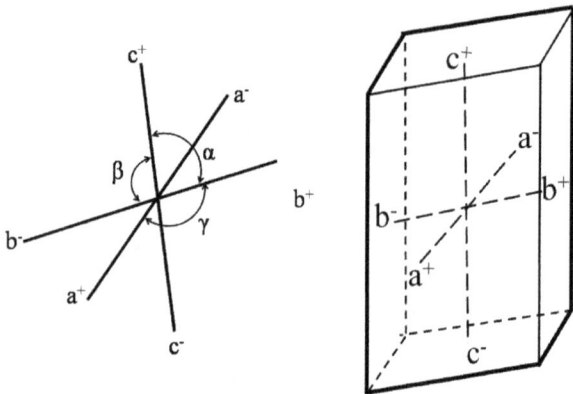

Figure 35: The crystallographic axes of the triclinic system

The Triclinic system has 2 classes as follows:
a. The Asymmetric (Triclinic pedial) class
b. The Triclinic holosymmetric (Triclinic pinacoidal) class, which is the holosymmetric class with the following properties:
  - A centre of symmetry
  - No planes of symmetry
  - No axes of symmetry
  - Examples include: *Plagioclase feldspars and Kyanite.*

1.8 Hands-on activities for Crystallography

Although Crystallography and most of its concepts can only be fully understood through practical activities and with the help of practical manuals and crystal models; the following are a few activities that can facilitate the understanding of some of its concepts.

Activity 1
- Aim: Understanding the concept of crystallisation, crystal faces and unit cells.
- Requirements: 8 square objects (for example; perfect maggi cubes, empty chalk boxes or perfect cubes of savon)
- Procedure: stack –up (pack) the 8 cubes as close together as possible to form a bigger cube (4 cubes on 4 cubes)

Questions;
1. How many flat surfaces surround each of your smaller cubes?
2. How many flat surfaces surround the bigger cube?
3. The process of packing or stacking up the smaller cubes into a bigger cube represents which natural process in crystallography?
4. What crystallographic name is given to the smaller cubes that make up the bigger cube?
5. If you add another layer of 4 cubes on the previous cube, what crystallographic name is given to the new form?

Activity 2
Aim: *Understanding the concept of planes of symmetry.*

-Requirements: 8 square objects (for example; perfect maggi cubes, empty chalk boxes or perfect cubes of savon)

Procedure: stack –up (pack) the 8 cubes as close together as possible to form a bigger cube (4 cubes on 4 cubes)

Questions;
1. In how many ways can you separate the bigger cube into two separate identical groups of 4 smaller cubes?
2. In crystallography, how do we call those lines along which the bigger cube is able to separate into two identical parts of 4 smaller cubes?

Activity 3
◊ Aim: Understanding common external features of a crystal.
◊ Requirements: A room at home or a classroom.
◊ Procedure: Study any part of the room where two walls meet and where two walls and the floor meet.

Questions;
1. What name is given to the angle formed where the two walls meet with the floor, if the room represented a crystal?
2. If two adjacent walls of a building represented crystal faces, what crystallographic term is given to the line formed when they meet?

Activity 4
◊ Aim: Understanding Common crystal forms.
◊ Requirements: Buildings and church bell towers (or palace buildings).
◊ Procedure: Study the roofing styles of school buildings, church bell towers and palace buildings

Question;
1. If the sides of the roof where crystal faces, what different crystal forms will they represent?

Bonus question
If a cube of sugar was a natural crystal, how will you describe its crystallographic axes? Under which crystal system will you classify it?

Study questions on crystallography
1. Define the following terms as used in crystallography;
   * Crystallography,
   * Unit cell
   * Amorphous,
   * Crystallisation,
   * Crystalline,
   * Anhedral,
   * Subhedral,
   * Euhedral,
   * Habit,
   * Interfacial angles,
   * Form,
   * Unlike faces,
   * Face,
   * Like faces,
   * Zone,
   * Edge,
   * Node,
   * Crystallographic axis,
   * Zone axis,
   * Symmetry,
   * Parameters,
   * Miller indices,
   * Intercepts,
   * Axis of symmetry pinacoids,
   * Plane of symmetry,
   * Centre of symmetry,
   * Domes
   * Prisms,
   * Pyramids,
   * Holosymmetric class.
   * Crystals,

2. Differentiate the following pairs of terms;
   - Open and closed forms
   - Like and unlike faces
   - Axis of symmetry and crystallographic axis
   - Dome and pyramid
   - Pinacoid and prism
3. Explain the law of constancy of interfacial angles.
4. Describe the crystallographic axes in the various crystal systems.
5. Outline the bases for the classification of crystals into systems and classes.
6. List the crystal systems in order of increasing abundance of classes.
7. For each of the crystal systems listed above;
   - State the characteristics of their crystallographic angles
   - The characteristics of their crystallographic axes
   - Their diagnostic property
   - Their holosymmetric classes and
   - An example of mineral that represents it.
8. Index all the faces on the following crystal models and identify each model, stating their systems;

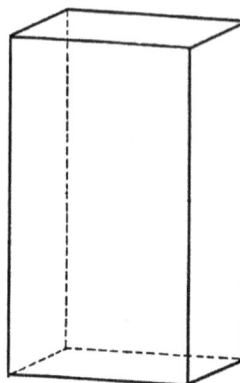

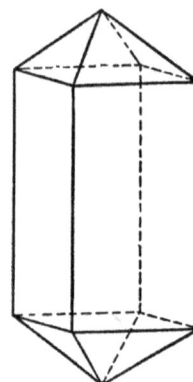

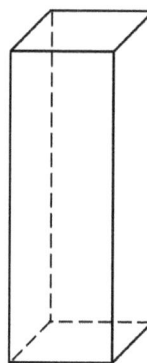

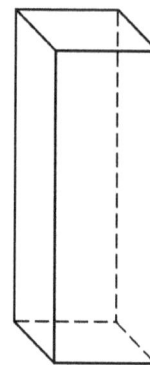

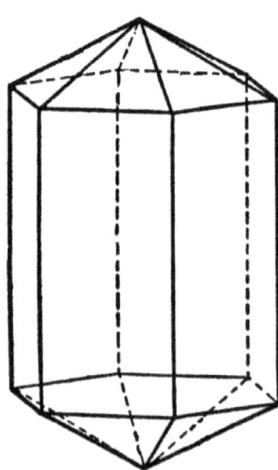

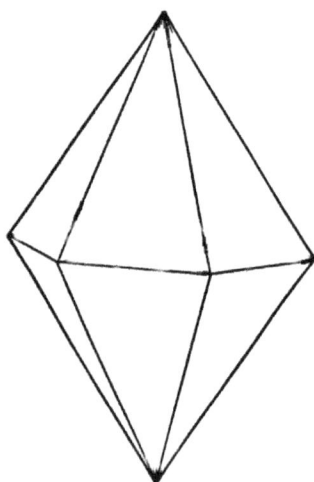

MINERALOGY

CHAPTER 2

Mineralogy

Chapter objectives

At the end of this chapter, you will be able to;

- Define minerals,

Explain:

- The origin of minerals
- The different types of bonding in minerals and state the effects of each type of bond on the physical properties of the mineral.
- Describe the main physical properties of minerals.
- Define isomorphism, polymorphism and pseudomorphism.
- Classify minerals into silicates and non-silicates based on chemical composition.
- Classify the silicate minerals based on atomic structure
- Explain the crystallography, physical properties and occurrence of the different groups of silicate minerals.
- Explain the effects of the structure of silicate minerals on their physical properties.
- Classify the non-silicate minerals based on their ionic composition and the radical present.

2.1 Introduction

Mineralogy is the study of minerals. These studies help mineralogists to identify, differentiate, classify and locate minerals. Minerals are of fundamental importance to geologists because it forms the basic constituent of rocks and can also be mined for economic purposes.

2.1.1 Definition

A mineral is a _naturally occurring_ homogenous solid with a definite (but not fixed) chemical composition and having a highly ordered atomic arrangement.

A close analysis of the underlined words in the above definition, clarifies the confusion between minerals and other economic resources that are often generally mistaken and called minerals.

    i. The words naturally occurring, distinguishes natural substances (formed by natural processes) like diamonds and emeralds from those synthesized in laboratories.

ii. The words underline{homogenous solid}, means that a mineral is composed of a single solid substance which cannot be physically subdivided into simpler chemical s diamorphic ubstances or compounds. It excludes gases and liquids though there are a few exceptions such as native mercury and natural gas which are termed mineraloids[1] . Water as ice is a mineral but not as a liquid or gas.

iii. The words *definite (but not always fixed) chemical composition,* imply that a mineral can be expressed by a specific chemical formula which may or may not be constant. Take for example the minerals Quartz and dolomite;

- Quartz is symbolised as SiO_2, therefore, any chemical substance with a compound other than Silicon dioxide cannot be Quartz. Quartz therefore has a definite and fixed chemical composition.

- Dolomite on the other hand, symbolised as $CaMg(CO_3)_2$, is not always pure Calcium (Ca), Magnesium (Mg) Carbon ate (CO_3). It may sometimes contain considerable amounts of Iron (Fe) and Manganese (Mn) in place of Magnesium (Mg). Because these amounts vary, Dolomite is therefore said to have a definite but not fixed chemical composition and can also be expressed as $Ca (Mg,Fe, Mn) (CO_3)_2$.

iv. The words *highly ordered atomic arrangement* indicate that minerals have an internal structure made up of a framework of atoms or ions arranged in a regular geometric pattern. This therefore causes minerals to be crystalline

solids and different from other solids such as glass or solidified candle wax which are described as amorphous.

Minerals are different from rocks in that minerals are compounds[2] composed of elements whereas rocks are an aggregate of minerals. For example the mineral Quartz, SiO_2, is a compound of the elements Silicon (Si) and Oxygen (O_2), whereas the rock granite is a mixture of several minerals, one of which is Quartz.

2.1.2 The Origin of Minerals and the Rock cycle

There are over 100 known elements in the Earth's crust. Unfortunately many of these are either rare, unstable and are of little importance to mineralogists.

Minerals are formed by the combination of two or more elements during natural processes operating within the Earth's crust. These processes include:

- Igneous processes (crystallisation from magma).
- Sedimentary processes (evaporation and precipitation by water).
- Metamorphic processes (re-crystallisation of pre-existing minerals under abnormal conditions of temperature and pressure).
- Surface processes (alteration of pre-existing minerals by weathering agents).

The natural processes that form minerals are the same processes that form and destroy rocks. The rock cycle is a model that describes the formation, breakdown, and reformation of a rock as a result of sedimentary, igneous, and metamorphic processes.

Sedimentary rocks can be formed from fragments of pre-existing igneous, metamorphic, or sedimentary rocks, or any combination thereof, or can be formed by organic or inorganic precipitation of common ions dissolved in salt or fresh water.

---

[1] Mineraloids are substances that resemble minerals in chemistry and occurance.

[2] Compounds are pure substances made up of two or more elements formed as a result of chemical change.

Igneous rocks can be formed through melting and extrusion or intrusion of pre-existing sedimentary, metamorphic, and/or igneous rocks.

Metamorphic rocks can be formed by alteration of pre-existing rock types through hydrothermalism, fault slip, or exposure to high temperatures and pressures due to deep burial.

The rock cycle operates in all geologic settings, ranging from the deepest part of the oceans (where oceanic seafloor reacts with seawater to form new "hydrated" minerals) to the highest peaks, where glaciation erodes and transports pre-existing rocks to lower elevations.

2.2 Bonding in Minerals

The combination of elements to form minerals occurs by bonding. Bonding depends on the valency of various atoms and their affinity for each other. The type of minerals to be formed from the crystallisation of magma depends also on the *atomic radius and electronegativity of the various atoms* in the melt.

Valency is the bonding potential of an atom, measured by the number of Hydrogen ions that the atom could combine with or replace.

In an ionic compound the valency (electrovalency) equals the ionic charge on each ion, for example in the compound MgO, Mg^{2+} shows a valency of 2, O^{2-} a valency of –2.

In a covalent compound the valency (covalency) of an atom is equal to the number of bonds it forms, for example in CH_4 Carbon has a valency of 4, Hydrogen a valency of 1.

There are four main ways by which elements can combine to form minerals namely:
- Ionic or heteropolar bonding,
- Covalent or homopolar bonding,
- Metallic bonding and
- Van der Waals (or residual) bonding.

Although certain minerals are characterised by certain types of bonds, two or more types may operate between atoms or groups of atoms in a single substance.

2.2.1 Ionic or Heteropolar bonding:

This type of bonding occurs between metal and non-metal atoms. There is a loss or gain of electrons by the constituent atoms so that they acquire either a positive or a negative charge (and are now called ions). The forces holding the ions together are those of electrical attraction between oppositely charged bodies. Each ion is surrounded by ions of opposite charge and the whole structure is neutral (that is with no overall charge).

For example: halite (sodium chloride) or common salt, (NaCl), made up of sodium and chlorine possesses ionic bonding.

Sodium has 1 electron on its outermost shell while chlorine has 7 electrons on its outermost shell. During bonding, sodium gives up one electron to chlorine for both atoms to attain a stable configuration.

Therefore sodium becomes a cation (positively charged) while chlorine becomes an anion (negatively charged).

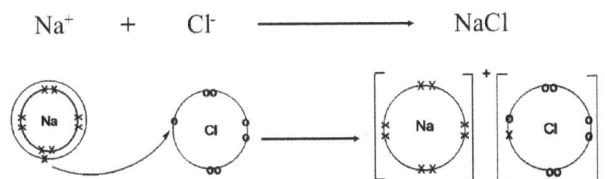

Figure 2.1: Ionic bonding process between sodium and chlorine.

Minerals with ionic bonds have the following properties;
- Most are brittle,
- Some are soluble.

- Examples include: Galena, Pyrite and Halite.

Covalent or homopolar bonding

In this type of bonding, the participating atoms contribute electrons which are mutually shared such that their outermost shells overlap.

For example: In Hydrogen, the single electrons are shared to make the Hydrogen molecule and in this way, the stable configuration of the nearest inert gas is achieved.

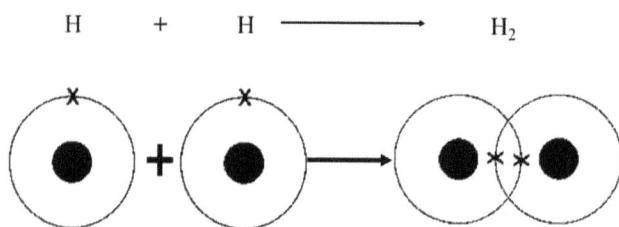

$$H \quad + \quad H \longrightarrow H_2$$

Figure 2.2: Formation of covalent bonds in a Hydrogen molecule.

Another example is in Oxygen where two atoms share two electrons in order to attain a stable configuration.

Diamond has a covalent structure composed entirely of Carbon atoms. The Carbon atom has four electrons in its outermost shell and thus can form four covalent bonds with other Carbon atoms. This is the basis of the diamond structure, in which each Carbon atom is surrounded by four others and this gives rise to a framework or network in 3 dimensions arranged as follows:

Because of the 3 dimensional frame-work in diamond, and because of the homopolarity of the bonds, diamond needs very high energy to break a bond so it is the hardest known substance on Earth. Covalently bonded minerals are also insoluble and are brittle, for example Quartz (SiO_2).

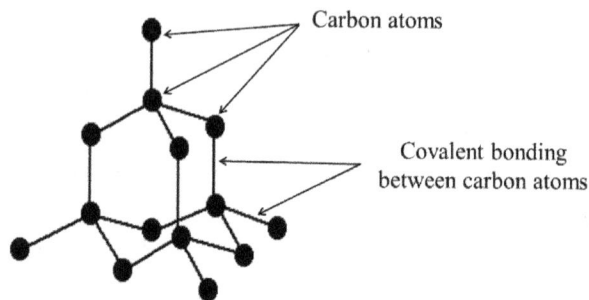

Figure 2.3: The structure of diamond

2.2.2 Metallic bonding

This type of bonding is exhibited by metals. It is the force holding metal cations together and a cloud of electrons. Pure metals or native metals are regarded as comprising of atoms with positively charged nuclei surrounded by a cloud of negatively charged electron shells which move freely and holding all metallic ions together.

This type of bonding makes metals to be good conductors of heat and electricity, having high ductility and malleability. For example in magnesium.

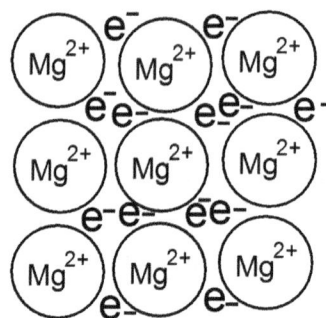

Figure 2.4: Metallic bonding in magnesium.

2.2.3 Van der Waals (or residual) bonding

This bond differs from all the others in that it is rarely responsible for the coherence of many common substances, although naphthalene is a notable exception. It is a weak force of attraction between the ions or atoms of all solids. It is less felt because in most substances its effects are shadowed by other types of bonding. The only solids entirely with Van der Waals bonding are inert gases (neon, argon etc...) in the solid state.

A good example is found in graphite which is a polymorph of Carbon . It is different from diamond in that the Carbon covalent bonds (intra-molecular bonds) result in the formation of hexagonal sheets instead of a framework and also, each Carbon atom is surrounded by only three others resulting in a deficiency of electrons. This deficiency is balanced by inter-sheet bonding (inter-molecular bonds) between respective hexagonal sheets of Carbon.

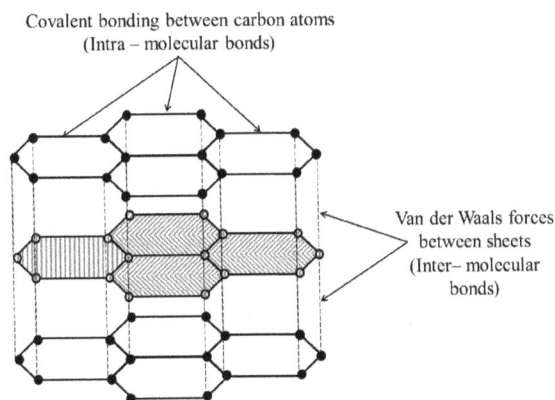

Figure 2.5: Van Der Waals forces in Graphite

This results in graphite having parallel sheets or layers which are linked to each other by weak forces or residual forces known as Van der Waals forces. These forces cause easy cleavage in graphite.

2.3 Physical Properties of Minerals

The physical properties of minerals are determined by its chemical composition and its crystalline structure. Since minerals have a definite chemical composition, most of their physical properties are also fixed and can be used to distinguish one from the other in hand specimen. The physical properties of minerals depend on a number of factors as follows;

- Physical properties or characteristics depending on light: colour, streak, lustre, transparency, translucency, phosphorescence and fluorescence.
- Physical properties or characteristics depending upon certain senses: taste, odour and feel.
- Physical properties or characteristics depending on the atomic structure and the state of aggregation (bonding): form, pseudomorphism, hardness, tenacity, fracture, cleavage and surface tension.
- Physical properties or characteristics depending on the specific gravity of minerals.
- Physical properties or characteristics depending on heat: fusibility and conductivity.
- Physical properties or characteristics depending on magnetism, electricity and radioactivity.

2.3.1 Physical properties or characteristics depending on light

a. **Colour**: It is the ability of a mineral to absorb and reflect some coloured rays (wavelengths) or vibrations of ordinary visible white light.

Therefore if a mineral looks blue, it means it has absorbed all the other vibrations of light and reflected only the blue vibrations. Similarly, if the mineral appears green, it means that it has absorbed all the other vibrations and reflected only the green.

When a mineral absorbs virtually all the wavelengths or vibrations and reflects virtually none, it appears black and when it reflects all the wavelengths and absorbs none, it appears white.

Colour is the most striking property of a mineral. However it is not very useful for mineral identification because:

Minerals belonging to the same group may

possess quite different colours. For example, in the Quartz group of minerals, Quartz (SiO_2) is usually colourless or white; but some of its varieties will appear brown, or pink or purple due to the amount of certain trace elements included in its composition. Also, two precious varieties of Beryl; emerald (green) and aquamarine (blue) exhibit different colours due to slight variations in its component trace elements.

There also exist situations where minerals show colour variations within a single crystal. Sometimes the crystals are arranged in a regular fashion, giving the mineral different colour bands (as in tourmaline) or the crystals are scattered in patches within the crystal (as in fluorite).

Factors influencing the colour of minerals

The true colour of a mineral depends on the nature and arrangement of its constituent mineral ions.

Minerals containing Aluminium (Al), Calcium (Ca), Sodium (Na), Potassium (K), Zirconium (Zr), Barium (Ba) and Strontium (Sr) as their main ions are generally light coloured or colourless for example; Calcite ($CaCO_3$), Barites ($CaSO_4$), Halite (NaCl) and Topaz $Al_2(SiO_4)(OHF)_2$, while those containing Iron (Fe), Titanium (Ti), Manganese (Mn), Chromium (Cr), Cobalt (Co), Nickel (Ni), Vanadium (V) and Copper (Cu) as their main ions are usually coloured or deeply coloured for example; Pyrite (FeS_2) and Chalcopyrite ($CuFeS_2$) which appear brass yellow, Magnetite (Fe_3O_4) appears iron black, Malachite ($CuCO_3(OH)_2$) appearing bright green. Most of the ferromagnesian minerals (iron/magnesium-bearing), such as augite, Hornblende, Olivine, and Biotite, are either green or black.

Different types of bonding can also cause minerals with the same chemical composition to have different colours for example in the allotropes of Carbon , diamond appears colourless and graphite appears black due to the different types of bonding of Carbon atoms in them.

The valency of an ion can affect the colour of a mineral for example minerals with Fe^{2+} are usually green, while those with Fe^{3+} are commonly red, yellow or brown. If both Fe^{2+} and Fe^{3+} are present in appreciable quantities, the mineral is usually black because of charge-transfer effects.

Some minerals when rotated or observed from different directions display a changing series of prismatic colours (also called a *play of colours*) similar to those seen in a rainbow or when looking through a glass prism. This is best shown by diamond, Quartz (and other colourless minerals) and certain varieties of plagioclase feldspars. It is produced by the splitting up of a ray of white light into its constituents as it enters and leaves the mineral.

Examples of *play of colours* include;
i.   **Schiller** which is common in hypersthenes, caused by reflection of colours on mineral plates exsolved on parallel planes within the crystal.
ii. **Iridescence**, due to interference of rays of light either by minute globules of water trapped in outer layers of a crystal lattice (also called *opalescence* common in opal) or by distortion in the atomic as in Labradorite lattice.
iii. **Tarnish** effects which appear on the surface of some minerals when they are exposed to air (and may also exhibit iridescence) is caused either by oxidation or the chemical action of sulphur and other agents present in the atmosphere. Tarnish can be distinguished from the true colour of the mineral by chipping or scratching the mineral. For example Chalcopyrite ($CuFeS_2$) and Bornite (Cu_5FeS_4) often tarnish to an iridescent mixture of colours and are sometimes referred to as "peacock ores".
•   **Pleochroism**, describes a phenomenon whereby some crystals display different

colours when rotated in plane-polarised transmitted light under a microscope. It is common in coloured minerals for example thin longitudinal sections of biotite display pleochroism.

The colour of minerals in the cubic system is usually constant throughout the minerals in all directions, for example: copper is copper red, galena is lead grey, graphite is black and magnetite is iron black. Minerals in other crystal systems often show a variation in colour, depending on the direction in which the mineral is viewed. Colour variation is sometimes also caused by impurities.

Colour is a diagnostic property for minerals with a constant and unique colour for example; azurite is deep blue, malachite is deep or bright green, iron is black, magnetite is iron black.

NB: When trying to obtain the colour of a mineral, the mineral should be dry and not weathered and observation should be done under natural light.

b. Streak: The streak of a mineral refers to the colour of its powder. It is obtained by rubbing or scratching the mineral across an unglazed porcelain plate also called a ***streak plate***. The streak of a mineral is more reliable in identification than colour because the powder of a mineral cannot be affected by impurities. Some minerals have streaks which are the same as the colour of the mineral in mass, while other minerals show streaks which are quite different from the mineral in mass (solid mineral). Minerals showing the same streak as the colour of the mineral in mass include:

Table 2.1: Minerals showing the same streak as the colour of the mineral in mass

Mineral	Colour	Streak
Malachite	Deep or bright green	Light green
Galena	Lead – grey	Grey
Azurite	Deep blue	Light blue
Magnetite	Iron black	Black

Minerals showing different streak from the colour of the mineral in mass include:

Table 2.2: Minerals showing different streak from the colour of the mineral in mass

Mineral	Colour	Streak
Augite	Black	White
Hornblende	Black	White
Pyrite	Brass yellow	White
Tourmaline	Appears as black, blue or yellow	White
Talc	Grey, brown or green	White

Laboratory work has proven that the use of streak in mineral differentiation can be effectively carried out with the Carbonates, phosphates, sulphides, oxides but not with the coloured sulphates because they often produce a white streak.

Based on streak characteristics, minerals can be divided or grouped into 5 categories as follows;

   i. No streak minerals which are harder than the streak plate due to very strong bonds for example diamond, garnet and corundum.

   ii. White or pale coloured minerals which produce white streaks for example Quartz, feldspars and calcite.

   iii. Black strongly coloured minerals which produce white streaks for example hornblende and augite.

   iv. Strongly coloured minerals which produce the same streak as the colour of the mineral in mass for example graphite, magnetite, azurite and galena.

v. Strongly coloured minerals which produce streaks which are not white but are also different from the colour of the mineral in mass for example limonite, chalcopyrite and pyrite.

c. **Lustre:** Defines the quality and intensity of light reflected from a mineral's surface; that is the way ordinary light is reflected from the surface of a mineral. Lustre is generally described qualitatively and comparative terms. Generally minerals with smooth surfaces tend to reflect more light than those with rough surfaces. Lustre can either be metallic or non-metallic, the following terms can be used to describe lustre;

i. **Metallic lustre:** It is the lustre of metals (it has the appearance of a metal).
- When weakly displayed, it is termed *submetallic*.
- When not displayed at all, it is termed *dull*.
- For example, gold, pyrite, iron and galena have a metallic lustre. Chromite and cuprite have a submetallic lustre and massive magnetite usually has a dull lustre.

ii. **Non-metallic lustre:** It is the lustre which is different from metallic lustre. It is very common among the silicates, Carbonates, sulphates, halides and non-metallic minerals.

iii. **Vitreous Lustre:** This is the lustre of broken glass.
- When well developed, it is termed subvitreous.
- When not developed at all, it is also called dull.
- For example, Quartz shows vitreous lustre, whereas the amphiboles (like Hornblende) and pyroxenes (like augite) are usually subvitreous or dull.

iv. **Resinous lustre:** This is the lustre of resin (like the appearance of tree gum like those from cypress trees or pine cones). It is well displayed by opal, amber sphalerite (a zinc ore) or some kinds of zinc blende.

v. **Pearly lustre:** This is the lustre of a pearl. It is shown by minerals with parallel faces or surfaces along which it can be separated into thin plates like a pile of thin glass sheets. Pearly lustre is shown by Talc, Brucite and Selenite.

vi. **Silky lustre:** This is the lustre of silk. It is common (peculiar) in minerals having a fibrous structure. The fibrous variety of gypsum known as satin – spar and the fibrous varieties of asbestos called amianthus are good examples of minerals with silky lustres.

vii. **Adamantine lustre:** This is the lustre of diamond. Crystals of the tin ore Cassiterite also exhibit this lustre.

viii. **Waxy lustre;** This type of lustre is shown by Serpentine.

ix. **Greasy lustre:** It is also known as oily lustre and is exhibited by minerals like graphite and talc.
- The lustre of minerals may be of ranging degrees of intensity, depending on the amount of light reflected from their surfaces.
- When the surface of a mineral is so brilliant that it reflects objects distinctly (like a mirror), it is said to be splendent.
- When the surface is less brilliant and objects are reflected indistinctly, it is described as being shinning.
- When the surface has no lustre at all, it is described as being dull.

d. Transparency and Translucency
An object is said to be transparent if it allows light to pass through with little or no interruption or distortion so that objects on the other side can be clearly seen.

Therefore, a mineral is transparent when the outline of an object seen through it is sharp and distinct; a clear Quartz crystal is a good example.

- A mineral is said to be **subtransparent** when an object seen through it is indistinct.
- Minerals which are capable of transmitting light but cannot be seen through are said to be **translucent**.
- If no light is transmitted, the mineral is said to be opaque (this refers to minerals in hand specimen).

Many opaque minerals, especially among the silicates become translucent when cut into very thin slices and can even become transparent when reduced to thin sections suitable for microscopic examination. Some other minerals, mostly oxides and sulphides such as Magnetite, Haematite, Ilmenite, Pyrite are always opaque even in thin sections.

e. Phosphorescence and Fluorescence

i. **Phosphorescence** is the property by which some substances emit light after having been subjected to certain conditions such as heating, rubbing, or exposure either radiation or to ultraviolet light.

For example;
- Some varieties of fluorite when powdered and heated on an iron plate, display a bright phosphorescence.
- When rubbed together in a dark room, pieces of Quartz emit a phosphorescent light.
- Many minerals also show phosphorescence when exposed to sunlight or even ordinary diffused light and quickly transferred to a dark room.
- Diamond, Ruby, Wilemite and a few other minerals exhibit a brilliant phosphorescence after exposure to X-rays. Some miners have used this property to make sure that all the ore containing the mineral (especially wilemite) has been extracted.

ii. **Fluorescence** is the property by which some minerals emit light when exposed to certain electrical radiations. This phenomenon is best displayed by fluorite, hence the name **fluorescence**.

2.3.2 Physical properties or characteristics depending upon certain senses: Taste, Odour and Feel.

a. Taste: This character is can only be perceived in minerals that are soluble in water. The following terms can be used;
  - *Saline* for the taste of common salt (NaCl or halite)
  - *Alkaline* for the taste of potash and soda (bases)
  - *Cooling* for the taste of nitre or potassium chlorate.
  - *Astringent* for the taste of green vitriol (hydrated iron sulphate)
  - *Sweetish astringent* for the taste of alum.
  - *Bitter* for the taste of Epsom salts (hydrous magnesium sulphate)
  - *Sour* for the taste of sulphuric acid.

b. **Odour:** Some minerals have characteristic odours when struck, rubbed, breathed upon or heated. Terms used to describe odour include:
  - *Alliaceous* (the odour of garlic); given off when arsenic compounds are heated
  - *Horse radish* (the odour of decaying horse-radish); given off when selenium compounds are heated.
  - *Sulphurous* (the odour of burning sulphur); given off by iron pyrite when struck or by many sulphides when heated.
  - *Fetid or foetid* (the odour of rotten eggs); given off by heating or rubbing certain varieties of Quartz or limestone.
  - *Argillaceous* or clayey; the odour of clay when

breath upon.

iii. **Feel:** This is the sensation a mineral gives which may help in its identification. It can be described as smooth, greasy or unctuous, harsh, meagre or rough. Certain minerals adhere to the tongue.

2.3.3 Physical properties or characteristics depending on the atomic structure and the state of aggregation (bonding):

a. **Hardness**

This is the degree of resistance a mineral offers to abrasion (scratching). The hardness of a mineral is very important in the physical identification of minerals and can be determined in many ways. Hardness may be tested by rubbing the specimen over a fine-cut file and noting the amount of powder and degree of noise produced in process. The less the powder and the greater the noise, the harder is the mineral. Soft minerals yield more powder and little noise. Another method used when comparing the hardness of two minerals is by rubbing the pointed edge of one of the minerals firmly across the flat surface of the other.

If the mineral with the pointed end is harder than the other mineral, it will scratch or leave a groove on the other mineral's flat surface. On the other hand, if it is softer, then it will leave no mark on the flat surface of the other mineral.

To determine the relative hardness of a mineral, a relative hardness scale is used. The most widely used hardness scale is the *Mohs's scale of Hardness* which was introduced by an Austrian mineralogist called *Moh*. On this scale, 10 minerals are designated as standards of hardness as shown on the table below:

Table 2.3: Mohs's Scale of hardness and the hardness of other common objects

Hardness	Standard mineral
1.	Talc
2.	Gypsum
3.	Calcite
4.	Fluorite
5.	Apatite
6.	Feldspar (K-feldspar)
7.	Quartz
8.	Topaz
9.	Corundum
10.	Diamond

The intervals on this scale are about equal, except for that between Corundum (9) and Diamond (10), which is estimated to be about ten times as great. That is if Corundum is 9, then Diamond is about 100 on the same scale.

On the scale, minerals with larger values can scratch those with lower values; that is Apatite (5) will scratch all the other minerals from Talc (1) to Fluorite (4) but not Feldspar (6) and Flourite (4) will also scratch all the other minerals with hardness values below its own but will not be able to scratch apatite (5).

On the field, the relative hardness of minerals can be determined using objects whose hardness is known. For example:
- Fingernail has a hardness of about 2.5, though this can vary from one individual to another.
- A copper coin has a hardness of about 3.5
- A steel knife has a hardness of about 5.1 - 6.5 depending on the quality of the steel.
- Glass has a hardness of about 5.5
- A streak plate has a hardness of about 6.5
- A file has a hardness of about 6.5

Precautions to take when determining the relative hardness of minerals

1. A definite scratch must be produced in the softer mineral. This is best seen by blowing away the powder (streak) produced by the scratching action and then examining the spot with a hand lens.
2. A softer mineral drawn across a harder one often produces a whitish stripe which may be mistaken for a scratch in the harder mineral.
3. Scratching a harder mineral with a knife blade produces a steel mark on them.
4. Granular specimens may give a kind of scratch by loosening and removing.
5. When carrying out a hardness test, use a fresh surface of the mineral and not one which is coated with decomposition products (weathered surface).

The hardness of a mineral obtained using the Mohs's scale or other materials is known as *the relative hardness.*

The absolute hardness of minerals is determined using a *Sclerometer,* which measures how much pressure is required to indent a mineral.

Factors affecting hardness

The hardness of a mineral depends on many factors, some of which include:

- The strength of the chemical bonding between the constituent ions, and
- The arrangement of the atoms affect the hardness of the mineral in that;

If the density, valency and bond strength increases, its hardness also increases.

Similarly, a mineral's hardness will decrease if the ionic size or radius increases.

Since hardness depends on the atomic structure of a mineral, the structural control results in a mineral's hardness varying in different directions on a crystal.

The difference is usually very small but in the mineral, Kyanite ($Al2SiO5$), which occurs as a bladed crystal it ranges between 7 and 5 depending on the direction.

b. **Tenacity:** This is a measure of how a mineral deforms when it is subjected to some form of deformation such as crushing or bending. A mineral's tenacity can be described as:

iii. **Sectile** : (Sectility); when the mineral can be cut with a knife and the resulting slice breaks under a hammer. For example graphite and gypsum

iv. **Malleable** (malleability): when the mineral can be cut into slices and the resulting slices can be hammered into thin flat sheets. For example native gold, silver and copper.

v. **Flexible** (flexibility): when a thin plate or lamina of the mineral remains bent after the force bending it has been removed. For example talc, chlorite and selenite.

vi. **Elastic** (Elasticity): when a thin plate or lamina of the mineral returns to its original shape after the force bending it has been removed. For example, the micas.

vii. **Brittle** (Brittleness): when the mineral crumbles or shatters easily. For example iron, pyrite, apatite and fluorite.

viii. **Ductile** (ductility): when the mineral can be drawn into thin wires. For example copper and aluminium.

c. **Cleavage:** This is the ability of a mineral to break or split along preferred or definite parallel planes producing smooth surfaces.

These definite planes are called *cleavage planes* and are closely related to the crystalline form and the atomic parking of the element.

In each mineral with cleavages, the directions of the cleavage planes are parallel to either a particular face or to a set of faces (representing a crystal form) in which the mineral will crystallise.

Minerals may show several cleavages, which are described by stating the crystallographic direction of each cleavage and also the degree of perfection of each cleavage plane.

Cleavage may be described in order of quality as perfect or eminent, good, distinct, poor, indistinct etc. Minerals can cleave along 1,2,3,4 and 6 planes.

- Minerals with 1 cleavage plane (usually having a sheet structure) include: Mica, Talc, Gypsum, Chlorite and Graphite.
- Minerals with 2 cleavage planes (cleaving in long fragments) include: Hornblende, Potash feldspars and Augite.
- Minerals with 3 cleavage planes (which intersect at right angles forming cubes) include: Halite and Galena
- Minerals with 3 cleavage planes (which intersect at oblique angles forming rhombohedrons) include: Calcite and Dolomite.
- Minerals with 4 cleavage planes (forming octahedrons) include: Diamond and Fluorite.
- Minerals with 6 cleavage planes (forming hexagons) include: Sphalerite.

Hornblende, amphibole and augite all have similar characteristics like colour and hardness, but can be distinguished through their angles of cleavage; augite has cleavage planes which intersect at about $87o - 90o$, while hornblende has cleavage planes which intersect at 56 degrees and 124 degrees.

The cubic system shows cubic cleavage (3 directions)

- The tetragonal system shows basal cleavage.
- The orthorhombic and hexagonal systems show basal and prismatic cleavage.

Certain rocks, such as slate, split readily into thin sheets and are said to be cleaved, but this property know as slaty cleavage is the result of recrystallisation of the rock under pressure and consequent mineral re-alignment and has no connection with mineral cleavage.

Glidding planes and 2 degree twinnings are related to cleavage.

d. Fracture:

This refers to the tendency for a mineral to break or chip producing an irregular surface totally independent of cleavage. Fracture is an important diagnostic character and a new one reveals the true colour of a mineral. The following terms can be used to describe fracture;

i. **Conchoidal fracture:** this is when the mineral breaks with a curved or convex fracture. This often shows concentric and gradually diminishing undulations towards the point of impact (resembling the growth lines on a shell). Examples of minerals with such a fracture include Quartz, flint and natural rock glasses particularly obsidian.

ii. **Even fracture:** When the fracture surface is flattish, as in chert.

iii. **Uneven fracture:** When the fracture surface is rough due to minute elevations and depressions. Most minerals show this kind of fracture

iv. **Hackly fracture:** When the fracture surface is covered by sharp and jagged elevations as in cast iron when it is broken.

v. **Earthy fracture:** Such as the dull fracture surface of chalk.

e. Form

Minerals and crystals normally grow freely outward into the melt or solution from which it is formed. Under these conditions, they are neither obstructed by other solid matter nor hindered by a shortage of the constituents needed for growth and therefore they assume a definite crystal form;

- If the mineral occurs with well-developed

crystals it is said to be *crystallized*.

- If the mineral possesses no definite crystals, but a confused intergrowth of imperfect crystal grains, it is described as crystalline.
- If the mineral shows only traces of a crystalline structure, then it is said to be *cryptocrystalline*.
- If there is a complete absence of crystal structure, the mineral is said to be *amorphous*. This is common in natural rock glasses (such as obsidian) but rare in minerals.

f. Habit

This is the external shape of a mineral formed by the development of an individual crystal or an aggregate of crystals. It depends upon the conditions during formation.

For example, one environment might give rise to long needle-like crystals while another will produce short platy crystals.

It is quite possible for the same mineral to have several different habits. Descriptive terms for mineral habits are split into those for individual crystals and those for aggregates of crystals as follows:

Those for individual crystals;

- *Acicular habit*: fine needle-like crystals
- *Bladed habit*: shaped like a knife blade or lath-like. It is commonly displayed by kyanite.
- *Fibrous habit:* consisting of fine thread-like strands as commonly shown by satin-spar (a variety of gypsum) and asbestos.
- *Foliated or foliaceous habit:* consisting of thin and separate lamellae or leaves as is shown by the mica group minerals and other sheet silicates silicates.
- *Lamellar habit:* consisting of separable plates or leaves as in wollastonite.
- *Prismatic habit:* elongation of the crystal in one direction as in the feldspars, the pyroxenes and common hornblendes.
- *Reticulated habit*: crystals in a cross-mesh

pattern, like a net as in rutile needles found within crystals of Quartz.

- *Scaly habit:* in small plates as in tridymite.
- *Tabular habit:* broad, flat, thin crystals as in sanidine feldspar.

Those for crystal aggregates;

- *Amygdaloidal habit:* almond-shaped aggregates, common in the zeolites, in which the minerals occupy vesicles or gas holes in lava flows.
- *Botryoidal habit:* spherical aggregations resembling a bunch of grapes as in azurite.
- *Columnar habit:* With massive aggregates in slender columns as is seen in stalactites and stalagmites usually with the mineral calcite.
- *Concretionary and nodular habit:* With spherical, ellipsoidal or irregular masses, as in flint nodules.
- *Dendritic and arborescent habit:* With aggregates in tree-like or moss-like shapes, usually with the mineral being deposited in crevasses or narrow planes as with the dendrites of manganese oxide.
- *Granular habit:* with coarse or fine grains. Evenly sized granular aggregates of minerals such as olivine in the ultrabasic rock dunite, are often termed saccharoidal because of their resemblance to lumps of sugar.
- Lenticular habit: consisting flattened balls or pellets, shown by many concretionary and nodular minerals.
- *Mammilated habit:* with large mutually interfering spheroidal surfaces, as in malachite.
- Radiating or divergent habit: With fibres arranged around a central point as in barite.
- *Reniform habit:* which is kidney-shaped has the rounded outer surfaces of massive mineral aggregates resembling those of kidneys. For example, kidney iron-ore, a variety of

heamatite.

- *Stellate habit:* with fibres radiating from a centre to produce star-like shapes as in wavelite.
- *Wiry or filiform habit:* with thin wires, often twisted like the strands of a rope as in native silver and copper.

g. Pseudomorphism:

This is a phenomenon whereby a mineral assumes the outward form of another mineral. The mineral which assumes the form is known as a pseudomorph and can be formed in the following ways:

i. By investment or incrustation, produced by depositing a coating of one mineral on to the crystals of another; for example, a coating of Quartz on fluorite crystals. If the fluorite crystals are later on dissolved, the Quartz will remain with the cubic form which is that of fluorite.

ii. By infiltration, where an original mineral is completely dissolved leaving a cavity such that the cavity previously occupied by the original mineral is refilled by deposition of a new mineral by infiltrating solutions.

iii. By replacement, where there is the slow and simultaneous substitution of particles of new and different mineral matter for the original mineral particles which are removed by solution. There is no chemical reaction between the materials. For example the substitution of wood for silica to form petrified wood.

iv. By alteration wherein the crystal undergo a gradual chemical change leading to their composition becoming so altered that they are no longer the same mineral but possess the same forms; for example, the alteration of olivine to serpentine.

Pseudomorphs may often be recognised by a lack of sharpness in the edge of the crystals, while their surfaces usually have a dull and somewhat granular or Earthy appearance.

h. Polymorphism:

This is the existence of one chemical substance or mineral in more than one form or structure. The two or more forms of the same mineral are called polymorphs or polymorphic forms. They usually have the same chemical composition but different physical properties like colour, hardness, specific gravity etc.

This difference in physical properties is due to the fact that they are formed under different conditions of temperature and pressure leading to a different atomic arrangement.

Examples of polymorphic forms are:

i. Calcite and aragonite which both have $CaCO_3$ as their chemical composition.

While calcite shows a rhombohedral form (in the Trigonal system), aragonite falls in the orthorhombic system, and while calcite has a hardness of 3 and a specific gravity of 2.71, aragonite has a hardness of 3.5 and a specific gravity of 2.94.

ii. Another *diamorphic* pair is diamond and graphite.

Quartz shows a polymorphic pair with polymorphs like low Quartz, high Quartz, Tridymite, Crystobalite, Coesite and Stishovite.

i. Isomorphism:

It refers to a situation where different minerals possess the same structural layout (isostructuralism). If the minerals are able to form a solid solution series, they are said to be isomorphous for example albite and anorthite. Rock salt and fluorite both crystallise in cubes but are not said to be isomorphous because they do not form a solid solution series.

A *solid solution series* is a mineral made up of several isomorphous members which are limited or found between well-defined end members which have a specific chemical composition.

The composition of the members within the solid solution series varies from one end member to another.

It is called a *solid solution* because during crystallisation the solid crystals formed, remain at equilibrium with the residual liquid (magmatic melt) whenever the composition varies from one end member to another.

In isomorphous series, ions easily substitute each other because they have similar ionic radii and ionic shapes. For example, olivine exists as a solid solution with the end members being fayalite ($FeSiO_4$) and forsterite ($MgSiO_4$). Here, magnesium has almost the same size as iron and also the same charge. Therefore, during the formation of olivine, the magnesium rich forsterite is first formed so that the solution left is richer in iron, and the forsterite formed exists in equilibrium within this solution. The olivine formed has an equal ratio of Mg and Fe; $(Mg_{50}Fe_{50})SiO_4$. At the end, the residual solution is highly enriched in Fe (iron), which then crystallises at low temperatures to form fayalite.

Other examples of solid solution series exist in plagioclase feldspars (Anorthite, $CaAl_2Si_2O_5$ and Albite, $NaA_1Si_3O_8$).

Due to the gradual variation in the composition of minerals in an isomorphic group, there are usually variations too in the physical properties of minerals within the same group and from one end member to another. For example, in the olivine solution series, forsterite is denser than fayalite, while fayalite is deeply coloured than forsterite.

j. Surface tension

This is the adhesive force that exists between a mineral and a liquid, causing it to float on the liquid. Different minerals have different degrees of floatation on various liquids, and this has formed the basis on which ore minerals can be separated from their gangues. For example the surface tension between various metallic sulphides and a particular liquid is greater than that between the gangue minerals such as Quartz, calcite, etc and the same liquid.

In the original Elmore process, a paste of sulphide and gangue was mixed with oil and water and agitated; the oil separated into a layer above the water and carried the sulphides with it.

The same principle more or less, underlies the method of extracting diamonds from their Kimberlite matrix, by causing them to adhere to grease upon a moving belt. Through surface tension effects, the working of some mixed ores has become economically possible.

2.3.4 Physical properties or characteristics depending on the specific gravity of minerals:

The specific gravity (SG) of a body is the ratio of the weight of the body to that of an equal volume of water at 4°C (4°C is the maximum density of water) expressed as a number.

The SG of minerals depends on:
- The atomic weight of the elements within the mineral.
- The manner in which the atoms are packed together.

For example, in isostructural compounds in which the atomic packing is the same, the compounds with atoms having higher weights have higher SG; An example is Olivine, having forsterite Mg_2SiO_4 (SG = 3.3) and fayalite Fe_2SiO_4 (SG = 4.4). Iron (Fe) has a higher atomic weight than magnesium (Mg).

Similarly, in polymorphous series like Carbon (with diamond and graphite), the difference in SG is controlled by an atomic parking. For example, diamond's SG is 2.54 due to the closely packed nature of its atoms, whereas graphite whose atoms are loosely packed has a SG of 2.3.

SG is of great importance in the identification of minerals.

The main principle employed in most

determinations of the SG of is that the loss in weight of a body immersed in water is the weight of a volume of water equal to that of the body.

(simply put as;. the weight of the volume of water displaced by a body is equal to the weight of the body...)

The body (mineral) is first weighed in air and then in water;

If **Wa** is the weight of a body in air, and **Ww** is the weight of the body in water,

Then **Wa – Ww** is the weight of water displaced by the body.

SG is calculated as follows:

$$SG = \frac{Wa}{Wa-Ww}$$

Methods of determining Specific Gravity (SG)

The methods of determining specific gravities in mineralogy are chosen depending usually on the *size and character* of the specimen under examination.

i. The normal chemical balance is used for fragments of a solid mineral about the size of a walnut (or the local cashew nut).
ii. Walker's Steelyard apparatus is used for large specimens.
iii. Jolly's spring balance is used for very small specimens.
iv. The pycnometer or specific gravity bottle is used for friable (powdery or crumbly) mineral grains or liquids.
v. By measuring the displaced water. This is suitable in obtaining the approximate SG for a number of specimens of a mineral.
vi. Heavy liquids are used mainly for the separation of mineral mixtures into their pure components according to their specific gravities and also for determining the approximate SG of mineral grains. The liquids most suitable for normal separations are;

– Bromoform or tetrabromoethane ($CHBr_3$), with a SG of 2.89,
– Methylene iodide with a SG of 3.3,
– Clerici's solution with a SG of 4.2.

Clerici's solution is not recommended nowadays because it is quite poisonous and can be absorbed by the body through the skin. Many other heavy liquids are available, but these are frequently poisonous and difficult to use.

Other oily solutions used include: **Mercurous Nitrate**, SG = 4.3, and **Thallium Silver Nitrate**, SG = 4.6, but they react with some sulphate minerals such as gypsum and they are also poisonous.

2.3.5 Physical properties or characteristics depending on heat: fusibility and conductivity.

The fusibility of some minerals is a useful character as an aid in their determination by the blowpipe. A scale of six minerals, of which the temperature of fusion was supposed to increase by equal steps was suggested by Von Kobell. These minerals and their approximate temperature of fusion are;

– Stibnite (525°C),
– Natrolite (965°C),
– Almandine garnet (1050°C)
– Actinolite (1200°C)
– Plagioclase (1300°C) and
– Olivine (1400°C)

2.3.6 Physical properties or characteristics depending on magnetism, electricity and radioactivity.
a. Magnetism;

This refers to the ability of a mineral to be attracted by an ordinary bar magnet or an electromagnet. Magnetite (Fe_3O_4) and to a lesser extent Pyrrhotite (Fe1-4S), are the only minerals affected by an ordinary bar magnet, but a larger number of minerals are affected by the electromagnet. A mineral containing iron does not necessarily mean that it should be magnetic, also the amount of iron in a mineral does not control its level of magnetism. For example monazite and other cerium bearing minerals which contain no iron are sufficiently magnetic to be electromagnetically separated from other non – magnetic minerals.

Electromagnetic separation is an important ore – dressing process. By varying the power of the electromagnet, minerals of varying magnetism can be separated from one another. For example the separation of magnetite from apatite, pyrite and siderite from blende, wolframite from cassiterite, and monazite from magnetite and garnet). Some ores are sometimes roasted, in order to boost their magnetism; For example pyrite and siderite. In the laboratory, small electromagnets have been used to separate heavy mineral residues from heavy liquids and also magnetic minerals from non-magnetic ones.

The magnetic properties of some common minerals are as follows:

- **Highly magnetic**: magnetite and pyrrhotite.
- **Moderately magnetic**: siderite, almandine, wolframite, ilmenite, hematite and chromite.
- **Weakly magnetic**: Tourmaline, spinel group, monazite
- **Non–magnetic**: Quartz, calcite, feldspars, corundum, blende or sphalerite, Cassiterite.

Palaeomagnetism is the study of past or ancient magnetic field of the Earth as preserved in certain rocks like basalts. When certain rocks crystallize, the magnetic particles in them become oriented in the Earth's magnetic field existing at that particular time and place. The directions of earlier magnetic fields and the positions occupied by the North pole at these times can be determined on rock samples. This has been of great importance in studying the movement of the large Earth plates during geologic time. (Plate tectonics).

b. Electricity

Minerals vary in their capacity to conduct electricity. Good conductors are relatively few in number and include those minerals with a metallic lustre such as native metals and sulphides (except sphalerite or zinc blende) which has an oily lustre.

Good conductors can be separated in the laboratory from bad ones by their attraction to a glass rod which possesses an electrostatic charge, preciously induced by rubbing the rod with a silk cloth.

- Metals with pure metallic bonding are excellent conductors. For example copper.
- Minerals with partial metallic bonding are semi-conductors. For example the sulphides.
- Covalently bonded minerals are usually non-conductors.
- Certain minerals develop an electric charge when subjected to certain conditions such as temperature change (pyroelectric minerals for example, tourmaline) or to a change in stress (piezoelectric minerals for example Quartz).

c. Radioactivity

Many minerals contain elements which are subject to radioactive decay throughout geologic time. Radioactive minerals are rare commonly occurring as salts of uranium and thorium and can be detected in the field using a *Geiger counter or a scintillometer*. However, many minerals contain certain radioactive isotopes which are subject to decay when the unstable parent isotopes break into stable daughter products. The time taken for half the original unstable parent isotope to breakdown (or decay) into stable daughter isotopes is called the *half-life*. The process of decay has been used to determine the age (radiometric age) of minerals using the following equation:

$t = (1/\lambda) \log_e [(D/P) + 1]$

Where:

- t is the age,
- λ is the decay constant,
- D is the concentration of daughter atoms in the mineral or rock due to radioactive decay

and
- P is the concentration of parent atoms in the mineral or rock.

A mass spectrometer is used to analyse for the stable and unstable isotopes in a mineral. For example the parent – daughter pair samarium (147Sa) and neodymium (143Nd) have recently been used with great success to determine the age of Precambrian rocks greater than 3000 million years old.

2.4 Occurrence and Classification of Rock Forming Minerals

2.4.1 Occurrence of minerals

Minerals are made up of elements or compounds, while rocks are made up of minerals. Over 100 elements are known today, but many of these are either rare or unstable and are of little importance to the mineralogist. Estimates of the chemical composition of the Earth's crust are based on many chemical analysis of the rocks exposed on the Earth's surface. The estimated percentages of minerals in the Earth's crust are shown in the table below:

Table 2.4: The estimated percentage composition and crustal abundance of elements in the Earth's crust

S/N	Element	Symbol	Percentage by weight	Percentage by volume
	Oxygen	O	46.6	93.8
	Silicon	Si	27.7	0.9
	Aluminium	Al	8.1	0.8
	Iron	Fe	5.0	0.5
	Calcium	Ca	3.6	1.0
	Sodium	Na	2.8	1.2
	Potassium	K	2.6	1.5
	Magnesium	Mg	2.1	0.3
	All others		1.5	-

From the above table, it can be seen that more than 98% of the Earth's crust is composed of just eight elements, out of which Oxygen is the most abundant (with 46.6% by weight). This percentage includes the Oxygen in the air we breathe, in water and in the rocks of the crust. By volume, Oxygen takes up a lot because its electron shells take up a large amount of space compared to its weight.

Silicon is the second most abundant element (with 27.7% by weight). It is the element which is used to make computer chips. Silica is a term for Oxygen combined with Silicon. Because of the abundance of Silicon and Oxygen, a majority of crustal minerals are silicates (compounds involving Silicon and Oxygen in the form of the silicate radical). For example Quartz (SiO_2) is pure silica that has crystallized.

Aluminium (with 8.1% by weight) is the third most abundant element, more common in rocks than iron. This should make aluminium less expensive than iron, but unfortunately this is not the case because aluminium is so strongly bonded to Oxygen and other elements that the procedures involved in breaking these bonds and extracting it is too costly for commercial production. Aluminium is therefore mined from uncommon deposits where aluminium-bearing rocks have been weathered, producing compounds in which the crystalline bonds are not so strong.

Other vital elements such as Hydrogen and Carbon do not feature on the above list because of their relatively low abundance. An element like copper which is highly used is 27th on the list of abundance but since the Earth's crust is not homogenous, geological processes are able to concentrate it into copper rich ores in certain locations that can be discovered and mined by exploration geologists. (These geological processes and the types of ores they produce are covered more under Economic Geology)

Nevertheless, the above 8 elements alongside the others combine in various proportions to form the different rock forming minerals of the crust.

2.4.2 Classification of rock forming minerals

The minerals of the crust can be classified based on several parameters but the most commonly used classification is that based on the composition of the radical present in the mineral. On this basis, the following groups of minerals are common:

Table 2.5: A classification of minerals based on the radical present in the mineral

S/N	Chemical group	Typical mineral	Formula
	Native elements	Gold	Au
		Silver	Ag
		Copper	Cu
	Non-metals	Carbon (diamond and graphite)	C
		Sulphur	S
	Oxides	Corundum	Al_2O_3
		Haematite	Fe_2O_3
		Ice	H_2O
	Hydroxides	Limonite	$Fe_2O_3.nH_2O$
		Brucite	$Mg(OH)_2$
	Sulphides	Pyrite	FeS_2
		Galena	PbS
		Chalcopyrite	$CuFeS_2$
	Halides	Halite	NaCl
		Fluorite	CaF_2
	Silicates	Feldspars	$NaCaAlSi_3O_8$
		Micas	$KAl_2(SiAl)$ $O.10(OH)_2$
	Carbonates	Calcite	$CaCO_3$
		Dolomite	$CaMg(CO_3)_2$
		Siderite	$FeCO_3$
	Sulphates	Gypsum	$CaSO_4.2H_2O$
		Barite	$BaSO_4$
	Phosphate	Apatite	$Ca_5(PO_4)_3(OH,FCl)$

The above minerals are embodied in the 3 rock types of the crust which are the igneous, metamorphic and sedimentary rocks.

As observed earlier, Oxygen, Silicon and aluminium are the 3 most abundant elements in the crust, combining to form a group of minerals called the silicates. Silicates are the most abundant, widespread and commonest rock forming minerals of the crust. They are formed by the combination of the silicate radical ($SiO4^{4-}$) with various cations and in various ways. Therefore minerals can be broadly divided into 2 groups; Silicates and non-silicate minerals.

2.5 The Silicate Group of Minerals

Silicates are an important group of rock forming minerals. This is because Oxygen and Silicon are the two most abundant elements of the crust which combine to form the basic building block for most common minerals. Silicates are formed when igneous magma crystallises.

2.5.1 The structure of silicate minerals

All silicate minerals contain silicate oxyanion $[SiO_4]^{4-}$ as its basic building block. In each building block, four Oxygen atoms are packed together around a single much smaller Silicon atom as shown below:

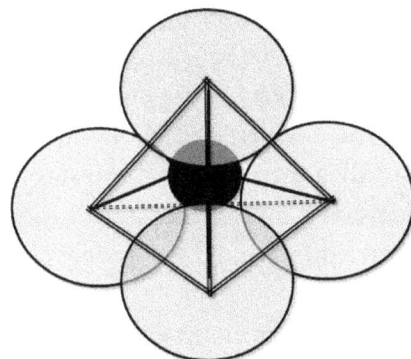

Figure 2.6a: The Silicon - Oxygen Tetrahedron

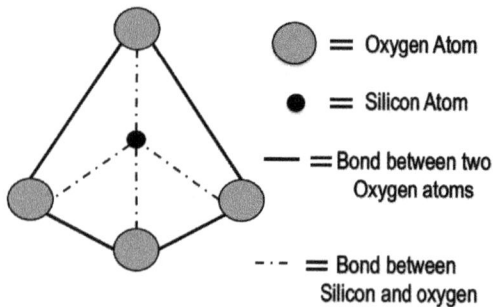

= Oxygen Atom

= Silicon Atom

= Bond between two Oxygen atoms

= Bond between Silicon and oxygen

The four-sided, pyramidal, geometric shape is called a *tetrahedron* and is used to represent the four Oxygen atoms surrounding a Silicon atom. This basic building block of a crystal is called a Silicon-Oxygen tetrahedron or a silica tetrahedron[3] .

The atoms of the tetrahedron are strongly bonded together. During this bonding, Silicon (which has +4 valency) shares its 4 outer electrons with the 4 Oxygen atoms surrounding it in order to obtain a stable configuration.

One of the rules for mineral formation is the attainment of electrical neutrality or balancing of various ionic charges within the mineral. Within a single silica tetrahedron, the negative charges exceed the positive charges, because Silicon has a charge of +4 and the four Oxygen ions have 8 negative charges all together (-2 for each Oxygen atom). This gives the tetrahedron a net charge of -4. That is $[SiO_4]^{4-}$.

For the silica tetrahedron to be stable within a crystal structure, it must either by;
- Balanced by enough positively charged ions or
- Share Oxygen atoms with adjacent tetrahedrons,

---

[3] silicon-oxygen tetrahedron or a silica tetrahedron is singular. The plural is silica tetrahedra or silica tetrahedrons.

And therefore reduce the number of positively charged ions required to balance it. Either way, both processes lead to a diversity seen among the silicates. Another cause of this diversity is brought about by the fact that in some silicates, some of the Silicon is substituted by other cations which usually have smaller ionic sizes but close to that of Silicon. For example; Fe^{3+} and Al^{3+}. This ionic substitution causes the silicate involved to have more charge deficits which require even more univalent cations to balance it up.

Some common cations which balance the excess charges are Ca^{2+}, Mg^{2+}, Na^+, K^+, Fe^{2+} and Cr^{3+}.

2.5.2 Classification of Silicate Minerals

Silicate minerals are classified based on the type of silica tetrahedron structure that makes up the mineral. The silica tetrahedron structure within the mineral depends on whether the silicate oxyanion $[SiO_4]^{4-}$ is balanced up by cations or by the sharing of all of its Oxygen atoms with other silica tetrahedrons thereby forming a framework. On this basis, 6 groups of silicates can be identified as follows:
- Nesosilicates,
- Sorosilicates,
- Cyclosilicates,
- Inosilicates,
- Phyllosilicates,
- Tectosilicates.

i. Nesosilicates (orthosilicates or island silicates):

This group of silicates contain discrete silica tetrahedrons with no Oxygen atoms shared.
- They are linked in the silicate structure by cations.
- Their standard formula is $[SiO_4]^{4-}$.
- They are usually balanced up by 2 divalent cations (like Mg^{2+} or Fe^{2+}).
- Examples include olivine $(Mg, Fe)_2SiO_4$, garnet $X_3Y_2[SiO_4]$ (where X may be Ca, Mg,

Fe^{2+}, or Mn and Y may be Al, Fe^{3+}, or Cr^{3+}), sphene, staurolite, topaz, etc.

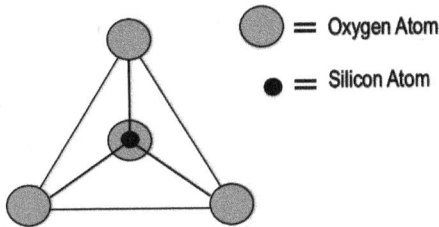

Figure 2.7: Nesosilicate structure.

ii. Sorosilicates:

This group of silicates contains groups of silicates sharing a common Oxygen atom forming pairs of silica tetrahedrons.

- Their standard formula is $[Si_2O_7]^{6-}$.
- The paired structure created requires 3 divalent cations to neutralize the net charge created (6^-).
- Examples are Melilite (Mg, Al) $(Ca,Na)_2$ $(Si,Al)_2 O_7$, Idocrase and Epidote.

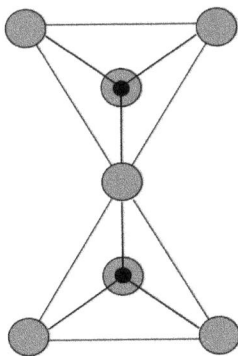

Figure 2.8: Sorosilicate structure.

iii. Cyclosilicates (ring silicates):

This group of silicates contains closed rings of 3, 4 or 6 silica tetrahedrons, with each tetrahedron sharing 2 of its Oxygen atoms.

Their standard formula is $[Si_nO_{3n}]^{2n-}$. Where n is a positive integer (whole number) indicating the number of tetrahedrons in the ring.

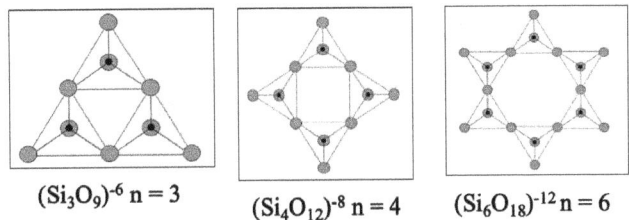

$(Si_3O_9)^{-6}$ n = 3 $(Si_4O_{12})^{-8}$ n = 4 $(Si_6O_{18})^{-12}$ n = 6

Figure 2.9: Varieties of Cyclosilicates.

A typical ring silicate has as its formula $[Si_6O_{18}]^{12-}$ For example:

- Beryl $(Be_3Al_2Si_6O_{18})$,
- Tourmaline Na $(Mg,Fe^{2+},Mn,Li,Al)_3Al_6(BO_3)_3[Si_6O_{18}](OH,F)_4$ and
- Cordierite $Al_3(Mg, Fe)_2[Si_5AlO_{18}]$.

iv. Inosilicates (chain silicates):

This group of silicates are formed when adjacent tetrahedrons share their Oxygen atoms and are linked up into straight continuous single chains which are joined laterally by cations or are stacked up into parallel double chains. Two types of inosilicates are the single chain and double chain inosilicates.

a. Single chain inosilicates:

In this group of inosilicates, the tetrahedrons share 2 Oxygen atoms each and are linked up into straight continuous single chains which are joined laterally by cations such as Magnesium (Mg^{2+}), calcium (Ca^{2+}) and Iron (Fe^{2+}) which lie between them.

The chains lie parallel to the c – axis.
- Their general formula is $[Si_2O_6]n^{4-}$.
- For example pyroxenes like Enstatite $Mg_2Si_2O_6$.

Since the Silicon bonds between the chains are stronger than the bonds holding separate chains, pyroxenes tend to split parallel to the chains in two directions which are nearly at right angles (88 degrees) to each other. The vertical cleavages in the mineral run between the chains.

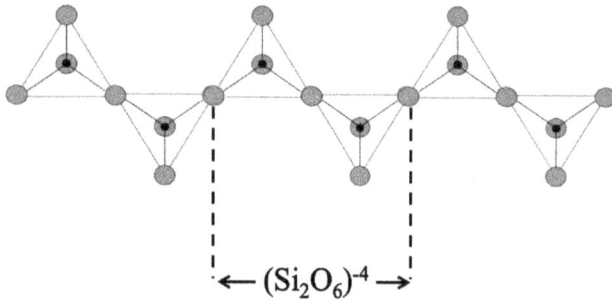

$\leftarrow (Si_2O_6)^{-4} \rightarrow$

Figure 2.10: Structure of Single chain Inosilicates.

In aluminous pyroxenes such as Augite, aluminium atoms may replace Silicon to a limited extent since they have nearly the same size. It may also occur among the atoms which lie between the chains.

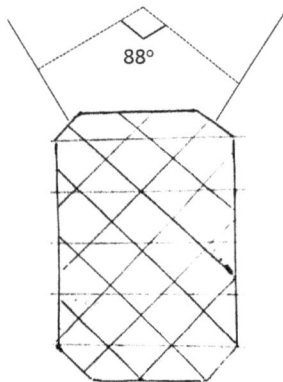

88°

Figure 2.11: 8 sided square shaped basal section of the pyroxenes

b. Double chain inosilicates:

In this group of inosilicates, 2 single chains tend to lie side by side with the tetrahedrons alternately sharing 2 or 3 Oxygen atoms. The resulting structure tends to be one Oxygen less than we find in single chains.
- The double chains are held together by monovalent, divalent, or trivalent cations, of which Na, Ca^{2+}, Mg^{2+}, Fe^{2+}, Al^{3+}, and Fe^{3+} are the most important. The hexagonal holes in the chains are occupied by hydroxyl ions (OH^-) to the extent of about 1:11 Oxygen.
- Their general formula is $[Si_4O_{11}]_n^{6-}$.
- The minerals of this group usually have a prismatic habit.
- For example amphiboles like the calcium-rich amphibole; Tremolite $(Ca_2(Mg,Fe_2)_5[Si_4O_{11}]_2(OH,F)_2$.

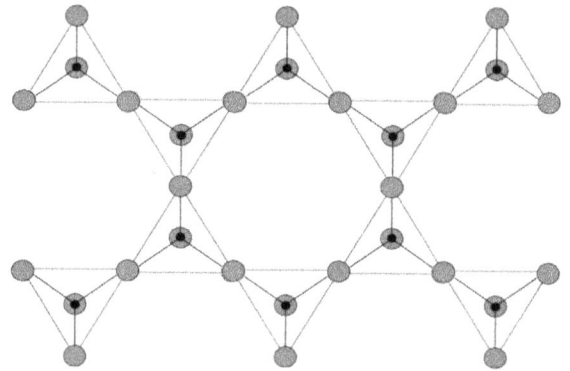

Figure 2.12: Structure of double chain Inosilicates.

Amphiboles are not strongly held together and so they have two pronounced cleavages at 124 degrees to each other.

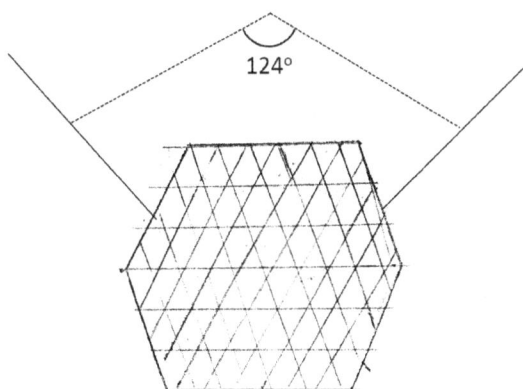

Figure 2.13: 6 sided diamon shaped basal section of the amphiboles.

In the amphiboles, aluminium can replace some of the Silicon to form aluminous amphiboles such as Hornblende (Si_4O_4 becomes $AlSi_3O_{11}$). This replacement requires the addition of some cations for it to be balanced. Aluminium (Al^{3+}) may be substituted for magnesium (Mg^{2+}) or ions like sodium (Na^+) being added. Ions of similar sizes can also change places. For example: iron (Fe^{2+}) or Manganese (Mg^{2+}) can substitute for magnesium (Mg^{2+}) and also iron (Fe2+), magnesium (Mg^{2+}) or two sodium ($2Na^+$) may substitute for calcium (Ca^{2+}).

Differences between Amphiboles and Pyroxenes

Table 2.6: Differences between amphiboles and pyroxenes

S/N	Parameter	Amphiboles	Pyroxenes
	Habit	Elongated prisms often acicular or bladed	Squat prisms (short prisms)
	Base cleavage	6 sided and diamond shaped basal section 2 prismatic cleavages at 124° in the basal section	8 sided and square shaped basal section 2 prismatic cleavages at 88° in basal section

S/N	Parameter	Amphiboles	Pyroxenes
	Pleochroism	Usually pleochroic	Non – pleochroic except for the Na – Fe^{3+} bearing varieties
	Composition	Hydroxylated	Unhydroxylated
	Structure	Double chains	Single chains
	Formula	$(Si_4O_{11})_n$	$(Si_2O_6)_n$

Similarities between Amphiboles and Pyroxenes

Many similarities exist between the pyroxenes and amphiboles including the following;

- Both groups have orthorhombic and monoclinic members
- The unit distance along the c and a directions are almost the same
- They contain the same cations but amphiboles are characterised by OH groups which is lacking in pyroxene.

v. Phyllosilicates (sheet silicates):

- This group of silicates result when three Oxygen atoms are shared between adjacent tetrahedra forming sheets.
- Their general formula is $[Si_4O_{10}]^{n \, 4-}$.
- The phyllosilicate sheets are stacked along the c axis with other sheets containing either brucite (with magnesium and iron ions) or gibbsite (with aluminium ions). Combinations of these sheets produce three different mineral types;

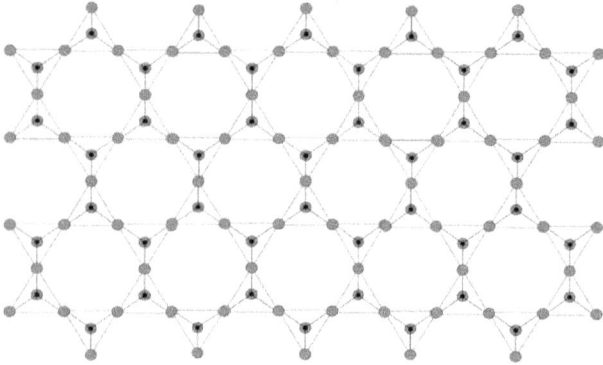

Figure 2.14: Structure of phyllosilicates.

- The 1:1 type made up of a two layer unit with one silicate sheet and a brucite or gibbsite sheet for example kaolinite and serpentine.
- The 2:1 type made up of a three-layer unit with a brucite or gibbsite sheet sandwiched between two silicate sheets for example micas, talc, Montmorillonite and all smectile clays.
- The 2:2 type made up of a four-layer unit with two silicate and two brucite or gibbsite sheets for example the chlorite group.

vi. Tectosilicates (framework silicates):

Tectosilicates are a silicate group of minerals in which all the 4 Oxygen atoms are shared with other tetrahedrons. This results to a 3 dimensional framework structure in which the SiO_4 is reduced to SiO_2. If the such a structure is composed entirely of Silicon and Oxygens, its composition is going to be $[SiO_2]_n$, and such is the composition of Quartz.

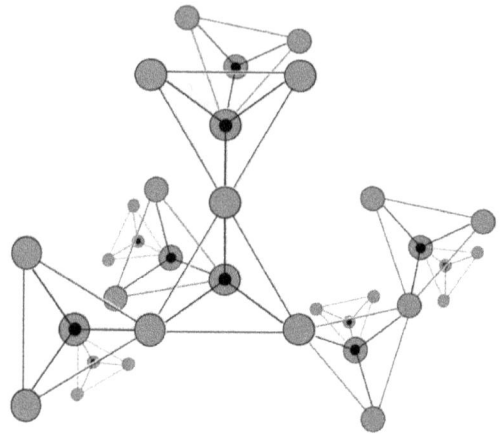

Figure 2.15: Structure of tectosilicates.

In this structure, the charges are balanced and there is no need for cation introduction.

However, sometimes there is the isomorphic substitution of some of the Silicon ions in the structure by aluminium (Al^{3+}). This substitution offers chances for univalent cations like sodium (Na^+) and potassium (K^+) and divalent cations like calcium (Ca^{2+}) to balance the charge deficit. This substitution brings about the diversity of the tectosilicates. Important tectosilicates include Quartz, the feldspars, feldsparthoids, zeolites and scapolite.

2.5.3 Properties of the silicates in relation to their structure (Internal atomic arrangement):

2.5.3.1 Physical Properties;

The complexity of the silicates increases from the nesosilicates down to tectosilicates. In this way the physical properties also vary from the independent silicate to the framework structure. This variation is due to the type of cations and the dense parking of the structure.

a. Density or Specific gravity;

This decreases with increase complexity of the

silicate structure. For example olivines and pyroxenes have an average specific gravity of 3.2 while feldspar and Quartz with more complex structures have an SG of 2.6. This is because in less complex structures, heavier ions are introduced and the atoms are more tightly packed than in complex silicates.

b. Cleavage

The atomic structure of the silicates influences the way they cleave. For example;

- Nesosilicates and sorosilicates generally have poorly developed cleavages if any.
- Cleavage in silicates is most pronounced in minerals whose internal structure have a regular pattern or is built along planes. These planes are occupied by cations which link major elements of the group such as in the pyroxenes and amphiboles which have 2 permanent cleavages along the planes where the tetrahedral chains are linked.
- The sheet silicates have a perfect basal cleavage which corresponds to the planes where the tetrahedral sheets are joined.

The effect of this type of structure in the sheet silicates is that most of the minerals in this group have the following:

- Platy habits
- One perfect basal cleavage
- Are generally soft
- Relatively low specific gravity
- Flexible or elastic laminae.

The one direction of cleavage corresponds to the plane where the potassium ions and Van der Waals forces bond the neighbouring sheets for example in muscovite.

On the other hand, the structures of tectosilicates do not have regular planes of weaknesses and their bonding is equal in all directions (evenly distributed) and this makes cleavage difficult in them.

c. Hardness

Hardness depends on the strength of chemical bonds within the mineral. The scratching of a mineral separates ions by breaking chemical bonds. Generally, less complex silicates are softer than the complex ones because a large proportion of their content is made up of cations linked by ionic bonds. If the bonds are weak such as in the case of Van der Waals forces or in ionic bonds, the degree of scratching will be high. For example the sheet silicates like talc, mica and chlorite are easily scratched by fingernail because the sheets in these minerals are held together by van der waals forces.

Most silicates however have high hardness (on Mohs's scale of hardness) especially the complex ones like Quartz. This is because of the strong covalent bond made by the silicate tetrahedron.

d. Refractive index

The refractive index decreases with increase in complexity of the silicates. For example, olivines and pyroxenes have a refractive index of between 1.7 – 1.8 while Quartz and feldspars have a refractive index of 1.4.

e. Temperature and time of formation:

The simple silicates like the olivenes are found to be formed early or at high temperatures from a magmatic melt while the more complex silicates like Quartz are formed late at low temperatures.

2.5.3.2 Chemical properties:

a. Ability to be hydrated:

The chain and sheet silicates have a high tendency to be hydrated (to store water) in the form of the hydroxyl radical $[OH]^-$. This is possible because of the open spaces and planes between the chains and sheets. For example, clay minerals are efficient water sponges.

b. Variable composition:

Many silicates display variable composition due to the diversity of the cations that are required to balance up the charge deficit in their silica tetrahedrons and also the ability of the cations to substitute each other within the silicate. This cationic substitution leads to the formation of isomorphic series (solid solution series) within some silicates for example in the nesosilicates, olivine has an isomorphic series between fayalite and forsterite caused by the substitution of magnesium in forsterite ($MgSiO_4$) for iron, giving rise to many intermediaries before ending with fayalite ($FeSiO_4$) which is the iron rich end member of the series.

Cationic substitution is possible when the cations involved have similar ionic radii and charges.

Table 2.7: Summary of the properties of the silicates

S/N	Name	Structural Group	General Formula	Example
	Nesosilicates	Discrete silica tetrahedrons with no Oxygen atoms shared.	$[SiO_4]^{4-}$.	Olivine: $(Mg, Fe)_2SiO_4$,
	Sorosilicates	Pairs of silica tetrahedrons sharing a common Oxygen atom.	$[Si_2O_7]^{6-}$.	Melilite: $(Mg,Al)(Ca,Na)_2 (Si,Al)_2 O_7$
	Cyclosilicates	Closed rings of 3, 4 or 6 silica tetrahedrons, with each tetrahedron sharing 2 of its Oxygen atoms.	$[Si_nO_{3n}]^{2n-}$. Where n is a positive integer.	Beryl $(Be_3Al_2Si_6O_{18})$
	Inosilicates	Single chain Inosilicates: Straight continuous single chains of tetrahedrons sharing 2 Oxygen atoms each.	$[Si_2O_6]^{4-}_n$ Where n is a positive integer.	Pyroxenes like Enstatite $Mg_2Si_2O_6$.
		Double chain inosilicates: 2 single side by side chains of tetrahedrons alternately sharing 2 or 3 Oxygen atoms each.	$[Si_4O_{11}]^{6-}_n$ Where n is a positive integer.	Amphiboles like tremolite (Ca$_2$(Mg,Fe)$_5$[Si$_4$O$_{11}$]$_2$(OH,F)$_2$.
	Phyllosilicates	Tetrahedrons form sheets by sharing 3 Oxygen atoms with adjacent neighbouring tetrahedrons.	$[Si_4O_{10}]^{4-}_n$ Where n is a positive integer.	Micas like Muscovite $K_2Al_4[Si_3AlO_{10}]_2(OH,F)_4$
	Tectosilicates	The four Oxygens in each tetrahedron are shared with other tetrahedra resulting in a framework structure.	$[SiO_2]_n$	Quartz; Si_O2

2.5.4 Other criteria for classifying silicates

Silicates (minerals that contain Silicon and Oxygen) can be sub-divided based on;
- Their cationic content and
- Their importance, abundance and time of formation in the rock in which they occur.

a. On the basis of cationic content, silicates can be sub-divided into;

- The *ferromagnesian minerals*, which contain iron (Fe) and magnesium (Mg) and
- The *non-ferromagnesian minerals*, which do not contain iron (Fe) and magnesium (Mg).

i. The ferromagnesian minerals are also known as mafic minerals and are usually dark in colour, very dense. They are also the first minerals to crystallise out at high temperatures from magma. Examples include pyroxenes, amphiboles and olivenes.

ii. The non-ferromagnesian minerals are also known as felsic minerals. They contain ions like calcium (Ca), sodium (Na) and potassium (K), which are light coloured and less dense compared ferromagnesian minerals. They usually crystallise later from magma at lower temperatures than ferromagnesian minerals. Examples include orthoclase, plagioclase and muscovite.

b. On the basis of their importance, abundance and time of formation, there are also two categories of silicate minerals namely;

- The *primary* minerals and
- The *secondary* minerals.

i. Primary minerals are those formed at the same time with the rocks that contain them for example olivines and pyroxenes in the igneous rocks like basalts. A primary mineral can either be essential or accessory.

– Essential primary minerals are those primary minerals whose presence is essential for the classification and naming of the rock, while

– Accessory primary minerals are those primary minerals whose presence does not affect the classification or naming of the rock.

ii. Secondary minerals are those formed as a result of the subsequent alteration of primary minerals. They are not important in naming or classifying the rock, but they help to determine the physico-chemical conditions of formation of the rock. Most secondary minerals are *hydrated silicates* (silicates formed as a result of the introduction of water into the rock system). A typical example is the alteration of primary olivine to secondary chlorite and serpentine.

2.5.5 Understanding Phase diagrams (Temperature/composition diagrams)

Phase diagrams are simply graphs that enable us to understand the temperatures and compositions of a magmatic melt at particular points (phases) during cooling.

These diagrams are constructed based on laboratory experiments such as those carried out by Bowen in 1922 which aimed at understanding the evolution of magmas as they crystallise.

A typical phase diagram such as those which represent the solid solution series in the plagioclases and the olivines looks as follows;

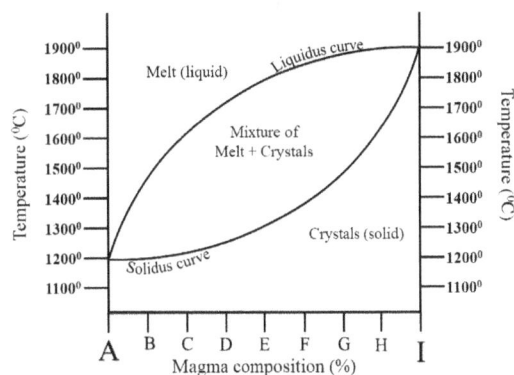

Figure 2.16: Parts of a phase diagram

From the diagram;

- The vertical axes represent temperature values,

while the horizontal axis represents the compositions scale of the members of the solid solution series. **A** and **I** are the end members while **B** – **H** represent the intermediate members of the solid solution series.

- The liquidus curve is the boundary which separates the liquid melt phase from the phase composed of a mixture of melt and crystals.
- The solidus curve is the boundary which separates the mixed crystals and melt phase from the solid crystals phase.

In summary;

- The portion above the liquidus is the liquid phase in which the magma melt is still completely molten with no crystallisation yet.
- The portion between the solidus and liquidus curves is a mixed phase in which the crystallization has begun. It contains a mixture of early formed crystals existing in equilibrium with the melt.
- The portion below the solidus is the phase in which crystallisation is complete. It is composed entirely of crystals.

2.5.6 Getting Compositions and Temperatures of formation from Phase Diagrams

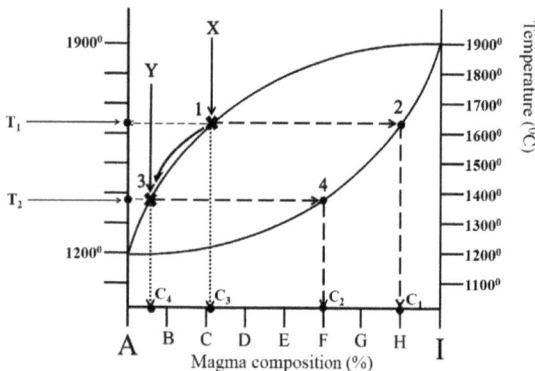

Figure 2.17: Reading temperatures and compositions on phase diagrams.

The above diagram demonstrates how a melt starts crystallising along the liquidus. The portion of the melt at X starts crystallising at point 1 and the portion at Y starts crystallising at point 3

The composition of the crystals (solids) in the melts at the points (1 and 3) is obtained by:

- Drawing horizontal lines from the points on the liquidus curve to the solidus curve (points 2 and 4 respectively), and then
- Drawing vertical lines from the solidus to the composition scale (points C_1 and C_2). The composition of crystals at point 1 is represented above by C_1 (which coincides with H) while the composition of crystals at point 3 is represented by C_2 (which coincides with F).

The temperatures at which crystallisation began can be obtained by drawing a horizontal line from the liquidus to the vertical axis. In the above diagram, the temperature of crystallisation of X is T_1, while the temperature for Y is T_2.

In summary;

- At point 1, melt X with a composition C_3 starts crystallising at a temperature T_1, producing crystals with composition C_1.
- Likewise at point 3, melt Y with a composition C_4 starts crystallising at a temperature T2, producing crystals with composition C_2.

Points to note

- C_3 and C_4 both indicate the compositions of the melts (liquid) at the various points; C_3 is the composition of the melt at point X while C_4 is the composition of the melt at point Y. C_1 and C_2 both indicate the compositions of the crystals (solids) in the melt.
- The crystals (under the solidus) have compositions which are always at equilibrium with the melt. For example;
- The crystals at point 2 are at equilibrium with

the melt at point 1 and
- The crystals at point 4 are at equilibrium with the melt at point 3.

Names of solid solution members	Composition	
	Magnesium (Mg) %	Iron (Fe) %
Ferrohortonolite	10%	90%
Fayalite	0%	100%

2.5.7 The Chemistry of Silicate Minerals

2.5.7.1 Nesolicates:

In this group, the silicate tetrahedral exists as ister or independent groups which are linked together by cations. Some of the Silicon ions are substituted by aluminium or iron. Common members of this group are Olivines and garnets, idocrase and Zircon.

a. The Olivines (Olivine groups)

The Olivine group of nesosilicates compose of an isomorphous series with the general formula R2SiO4 in which R is either Mg^{2+} or Fe^{2+}.

The magnesium end members is forsterite (Fo), Mg_2SiO_4 and the iron end member is Faylite (Fa) Fe_2SiO_4, so the general composition of Olivine is $(Mg,Fe)_2 SiO_4$ with Ni^{2+} and Co^{2+} Substituting for Mg^{2+} and Fe^{2+}.

Forsterite, crystallises at low temperatures, while fayalite crystallises at high temperature.

Between Faylite and Forsterite, there exists a solid solution series in which the Olivines are formed by the isomorphic substitution of magnesium and iron every time the temperatures are falling or rising.

Table 2.8: Names and compositions of the Olivene isomorphic series

Names of solid solution members	Composition	
	Magnesium (Mg) %	Iron (Fe) %
Forsterite	100%	0%
Chrysolite	90%	10%
Hyalosiderite	70%	30%
Hortonolite	30%	70%

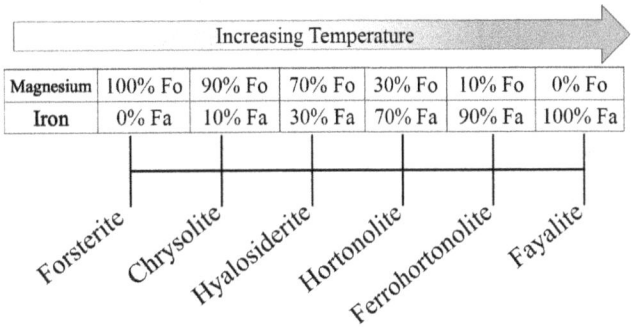

Figure 2.18: The Olivene isomorphic series.

Olivines are the first silicate minerals to be formed from a melt (magma). The following crystallisation curve shows how a solid solution arises during crystallisation.

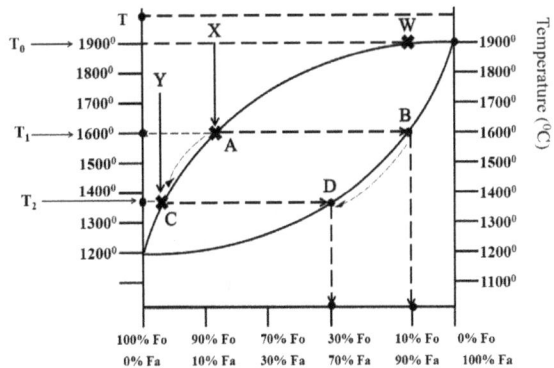

Figure 2.19: Phase diagram for fayalite – forsterite isomorphic series

From the above phase diagram of the forsterite – fayalite solid solution series;
- At temperature, T, no crystallisation is taking place.

- As the melt cools to 1900 °C (T_0), crystallisation begins. The composition of the crystals at point W is 100% fayalite and 0% forsterite; that is, the olivine crystals produced are richer in iron (Fe) with no magnesium (Mg).
- The melt at point X starts crystallising at point A, with crystals B which have a composition of 90% fayalite and 10% forsterite at a temperature of 1600 °C (T_1)

As the melt progressively cools, the portion at Y starts crystallising at point C, with temperature T2 (about 1380° C), producing crystals D which have a composition of 70% fayalite and 30% forsterite.

The phase diagram helps us to understand the fact that during the evolution (crystallisation) of a magmatic melt, the first olivine crystals, W, produced are richer in iron (fayalite) than in magnesium (forsterite) and as the cooling process progresses, the residual melt, Y, becomes more and more enriched in magnesium (forsterite). However, because the early formed crystals react continuously with the residual melt, the composition of the solid which finally crystallises has the same composition as the original melt.

Physical Properties

- *Crystal system:* Orthorhombic
- *Form/Habit:* Massive and compact grains but well-formed crystals are prismatic
- *Colour:* Shades of green, pale green, Olivine green, brownish or yellow in forsterite and brown or black in fayalite.
- *Lustre:* Vitreous
- *Hardness:* 6-7
- *Cleavage:* Poor on {010}
- *Fracture:* cracks sub-parallel to {001}
- SG: 3.22 (Fo) to 4.39 (Fa)
- *Diagnostic property:* Greenish colour
- *Occurrence:*

* Magnesium rich Olivine occurs as essential minierals in most ultrabasic igneous rocks like

Peridotite, Picrite and Dunites. These rocks are almost entirely composed of Olivine.

* It also occurs in basic igneous rocks like gabbro, dolerite and basalts. More Fe rich Olivine occur in intermediate rock

Olivines are rarely formed in association with silica (Quartz). This is mainly due to the difference in their temperatures of crystallisation; Olivine is the first mineral to be formed at about 1900°c while Quartz is the last to be formed at low temperatures of about 567°c. So by the time Quartz is crystallised in a melt, all olivine would have crystallised and sunk (precipitated) to the bottom of the magma in the chamber.

Zircon
Composition: $ZrSiO_4$
Physical Properties
- Crystal system: Tetragonal
- Form/Habit: Prismatic and pyramidal
- Lustre: Adamantine
- Colour: colourless, brown, grey, greenish.
- Fracture: Conchoidal
- Hardness: 7.5
- S.G: 4.6 – 4.7
- Occurrence: It occurs as an accessory mineral in igneous rocks such especially the more acidic types (granites) and pegmatites.

b. *The Garnet Group*

The garnet group of nesosilicates are a series of silicate minerals containing various divalent and trivalent cations with the general formula $X_3Y_2Si_3O_{12}$;

where X represents divalent cations like Ca^{2+}, Mg^{2+} and Fe^{2+}, Mn^{2+} and

Y represents trivalent cations notably Al^{3+}, Fe^{3+}, Cr^{3+} and Ti^{3+}.

The principal mineral in the garnet family are:
- Grossular $\quad Ca_3Al_2Si_3O_{12}$
- Pyrope. $\quad Mg_3Al_2Si_3O_{12}$

- Almandine $Fe_3Al_2Si_3O_{12}$
- Spessartine $Mn_3Al_2Si_3O_{12}$
- Andradite $Ca_3Fe_2Si_3O_{12}$
- Uvarovite $Ca_3Cr_2Si_2O_{12}$

Physical Properties

Crystal system: All garnets are cubic and fall in the hexaoctahedral class

- Form/Habit: Commonly show dodecahedron, or a combination of Tetrahexahedron and dodecahedron.
- Cleavage: No cleavage
- Hardness: 7 – 7.5
- Lustre: Vitreous
- Occurrence: Garnets occur in a variety of rock types;

* Almandine occurs in medium and high grade regional metamorphic rocks like schist and gneiss produced from metamorphism of pellitic rocks.

* Grossular and Andradite are common in metamorphosed impure limestones and skarns.

* Uvarovite occurs in serpentinites which are rich in chromites.

* Pyrope occurs in ultrabasic igneous rocks such as garnet peridotites.

Uses:

* Garnets are used abrasives

c. Aluminium Silicate Group (Al_2SiO_5):

Al_2SiO_5 exist as ploymorphs. There are 3 polymorphs of Al_2SiO_5 which are Andalusite, Sillimanite and Kyanite. The minerals contain individual silicate tetrahedral linked by chains of Al – O (Aluminium to Oxygen) groups. The different polymorphs have the same composition but different atomic structures and are stable under different physical conditions of temperature and pressure, as shown below;

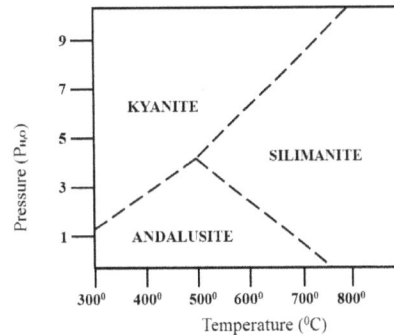

Figure 2.20: Pressure/composition diagram for the Aluminium silicates (Kyanite, Andalusite and silimanite)

Andalusite
Physical Properties

- Crystal system: Orthorhombic
- Colour: Pearl grey, purplish red
- Hardness: 6.5 – 7.5
- Fracture: Uneven, tough
- Cleavage: Poor {110} prismatic
- Lustre: Vitreous
- S.G: 3.1 – 3.16
- Occurrence: It occurs in thermally metamorphosed argillaceous rocks, forming andalusite hornfelses in the inner zones of thermal aureoles. It also occurs as an accessory mineral in certain granites.

Kyanite
Physical Properties

- Crystal system: Triclinic
- Form/Habit: Formed as long thin bladed crystals
- Colour: light blue and sometimes white and sometimes zoned from blue in the core of the crystal to colourless in the margins. Also grey green and rarely black.

- Cleavage: Good {100} and {110}: a parting on {001} is usually present.
- Lustre: Pearly for cleavage faces
- Hardness: 5.5 – 7.0
- S.G: 3.58 – 3.65
- Occurrence: It is typically found in regionally metamorphosed pellites under moderate heat flow and moderate to high pressure, forming kyanite schists and gneisses; sometimes found in eclogites.

Sillimanite

Physical Properties

- Crystal system: Orthorhombic
- Colour: Pearl grey, purplish red
- Cleavage: Poor {110} prismatic
- Form/Habit: Occur as long accicular crystals with diamond shaped cross sections.
- Lustre: Vitreous
- Hardness: 6.5 – 7.5
- Fracture: Uneven, tough
- S.G: 3.23 – 3.27
- Occurrence
  * Sillimanite occurs in hornfelses in the inner zones of thermal aureoles, resulting from the thermal metamorphism of pellitic rocks
  * It also occurs in high grade regionally metamorphosed argillaceous rocks.

Generally, the aluminosilicates occur as index minerals in metamorphosed aluminous rocks.

Uses of aluminium Silictes

* The aluminosilicates are mined in several parts of the world and used as refractory materials.

2.5.7.2 Sorosilicates

a. Epidote Group

The epidote group of minerals are complex silicates with the general formula $X_2Y_3Z_3O_{12}$ (OH,F)

Where :

- X = Ca, Ce^{3+}, La^{3+}, Y^{3+}, Fe^{2+}, Mn^{2+} and Mg^{2+}

- Y = Al^{3+}, Fe^{3+}, Mn^{2+}, Fe^{2+} and Ti^{3+}
- Z = Si^{4+}.

The formula can be written as $X_2Y_3[Si_2O_7][SiO_4]$ O(OH,F) showing that the atomic structure contains both $[Si_2O_7]$ and $[SiO_4]$ groups. Epidote minerals crystallise both in the Orthorhombic and Monoclinic crystal systems.

Minerals in this group are;

- Zoisite: $Ca_2Al_3Si_3O_{12}$ (OH)
- Clinozoisite $CaAl_3Si_3O_{12}$ with some Al^{3+} replaced by Fe^{3+}.
- Epidote: $Ca_2(Al2Fe^{3+})$ Si_3O_{12} (OH)

b. *Melilite Group* $(Ca, Na)_2(Mg, Al) [(Si, Al)_2O_7]$

The group comprises a series of minerals from a Ca – Al end member to a Ca – Mg end member.

The mineral Melilite occupies an intermediate composition with Ca, Al and Mg present. Na^+ and Fe^{3+} can replace Ca and Al respectively.

The minerals in this group are tetragonal and usually appear as small, tabular crystal or grains, white, yellowish or greenish in colour.

They are decomposed by HCl, with gelatinization. Melilite occurs in basic lava flows which are silica under saturated and without feldspar such as Melilite basalts.

2.5.7.3 Cyclosilicates

Beryl

Composition: $Be_3Al_2Si_6O_{18}$

Physical properties

- Crystal system: Hexagonal
- Form/Habit: Large Hexagonal prisms.
- Colour: Various shades of green, blue, yellowish or white
- Cleavage: Basal [0001] imperfect
- Fracture: Conchoidal or Uneven.
- Lustre: Vitreous or resinous

- Hardness: 7.5 – 8.0
- S.G: 2.66 – 2.92

Occurrence

* Beryl occurs as an accessory mineral in acid igneous rocks such as granite and pegmatite. It also occurs in metamorphic rocks of various types particularly schists.

Cordierite

Composition: $Al_3(Mg,Fe)_2(Si_5Al)O_{18}$

Physical properties
- Crystal system: Orthorhombic (pseudo-hexagonal)
- Form/Habit: Short pseudo-hexagonal crystals usually granular or massive
- Colour: Blue of various shades.
- Cleavage: 2 poor cleavages on {010} and {001}
- Fracture: sub-conchoidal.
- Lustre: Vitreous
- Hardness: 7
- Tenacity: Brittle
- S.G: 2.53 – 2.78

Occurrence

* Cordierite occurs mainly in metamorphic rocks; in regional metamorphic rocks, it occurs as cordierite gneisses or schists of high metamorphic grade and in thermal metamorphic rocks it occurs as cordierite hornfelses of pelitic composition.

* It is common in granites and granodiorite and their extrusive equivalent.

Uses

Used as gemstone. The precious variety is called water sapphire

Tourmaline

Composition – $Na(Mg, Fe^{3+}, Mn, Li, Al)_3, Al_6(BO_3)_3 Si_6O_{18}(OH, F)_4$. Fe_{3+} can replace Al in the lattice which contains $[Si_6O_{18}]$ rings.

Physical Properties

- Crystal system: Trigonal
- Form/Habit: Common as elongated prismatic crystals with a triangular cross section and sometimes acicular needles.
- Colour: Commonly black or bluish black, more rarely blue, pink or green and almost never colourless. The colours are often zoned with banding occurring along the prism length from top to bottom.
- Cleavage: A good prismatic cleavage {112 0̄} good and a poor rhombohedral cleavage {101̄0}.
- Fracture: Sub-conchoidal to uneven.
- Lustre: Vitreous
- Hardness: 7.0 – 7.5
- Tenacity: Brittle
- S.G: 2.9 – 3.2

Occurrence

* Tourmaline occurs in granite pegmatites, pneumatolytic veins and some granules. In the pneumatolytic stage, it is formed by alteration caused by Boron rich gases. Here, it is found in association with topaz, spondumene, fluorite, apatite and cassiterite.

* It also occurs in metamorphosed impure limestone.

* Tourmaline is a common detrital heavy mineral in sedimentary rocks and has been found as an authigenic mineral in some limestone.

-Uses : Some varieties are used as gemstones.

2.5.7.4 Inosilicates (Chain Silicates)

As earlier explained under the classification of silicate minerals, the inosilicates are silicates in which the SiO_4 are linked up persistently by sharing 2 or 3 Oxygen atoms. This linkage of the SiO_4 results to the formation of the chains structure. The inosilicates contain of 2 rock forming groups of minerals.

- Pyroxenes which exist as single chains and
- Amphiboles which exist as double chains.

i. Single chain inosilicate

a. Pyroxene group

- The pyroxene group includes both orthorhombic pyroxenes (orthopyroxenes) and monoclinic pyroxenes (clinopyroxenes).
- They all possess $[SiO_3]n$ in their structure.
- The general formula for pyroxenes is $X_{1-n}Y_{1+n}Z_2O_6$;

Where:

- X = Na or Ca,
- Y = Mg, Fe^{+2}, Ni, Li, Fe^{3+}, Cr or Ti and
- Z = Si, or Al.

In the orthopyroxenes, n = ~ 1 and thus x = 0 and the formula is reduced to $Y_2Z_2O_6$ or YZO_3, and virtually no monovalent or trivalent cations enter the structure except under high pressure when Al_{3+} may enter.

In the clinopyroxenes, n = 0 – 1 and the cations entering the structure must be such that the sum of charges in the X + Y = Z groups balance the six O_{2-} anions.

The pyroxenes are characterised by two prismatic cleavages which intersect almost at right angles on the basal plane.

A large number of clinopyroxenes (monoclinic pyroxenes) fall between a ternary system in pyroxenes, with Wollastonite (Wo) $CaSiO_3$, Enstatite (En) $MgSiO_3$, and Ferrosilite (Fs), $FeSiO_3$ as end members. The mid-points of two sides of the system represent diopside $(Ca_{0.5}Mg_{0.5})SiO_3$ and Hedenbergite $(Ca_{0.5}Fe_{0.5})SiO_3$, as shown below:

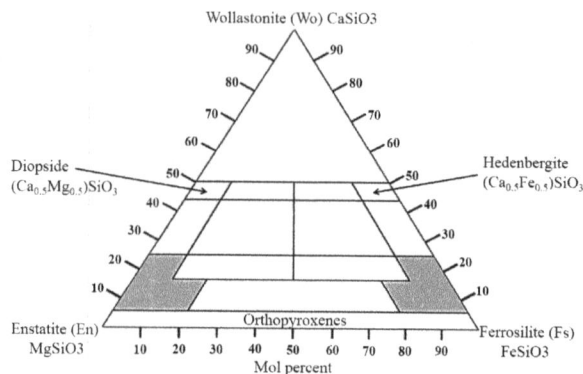

Figure 2.21: Composition diagram for pyroxenes indicating the positions of the end members, the mid-point members and the orthopyroxenes

The orthopyroxenes and clinopyroxenes show solid solution series as follows;

The Orthopyroxene series consists of the Enstatite $En_{100}Fe_0$ – Ferrosilite $Fe_{100}En_0$ with their intermediate members being Bronzite and Hypersthene

Occurrence of orthopyroxenes

- They occur in basic and ultrabasic igneous rocks of all types.
- Orthopyroxenes also occur in high grade regionally metamorphosed rocks (particularly Charnockites and Granulite) and in hornfels formed from pellitic rocks in the innermost zone of thermal aureoles
- Orthopyroxene commonly occur in some Chondritic meteorites.

The Clinopyroxenes are pyroxenes that crystallises in the monoclinic system. They are a diverse group of pyroxenes and show solid solution series such as;

- Diopside – Hedenbergite series. ($CaMg Si_2O_6 – CaFeSi_2O_6$)
- Augite – Ferroaugite series ($CaNa(Mg, Fe^{2+}Mn, Fe^{3+} Al, Ti)_2(Si,Al)_2O_6$
- Aegirine (Ae) – Aegerine augite series

$$(NaFe^{3+}Si_2O_6 - Na\ Ca\ (Fe^{2+}\ Fe^{3+}Mg)Si_2O_6$$

Occurrence of clinopyroxenes

- Diopside occurs in metamorphic rocks particularly metamorphosed dolomitic limestones and calcerous sedimentary rocks, and also in some basic extrusive igneous rocks.
- Hedenbergite occurs in metamorphosed iron – rich sedimentary rocks, and in some acid igneous rocks such as ferrogabbros and granophyres.
- Augites occur in igneous rocks and are essential constituents of gabbros, dolerites and basalts.
- Aegerine and aegerine augite occur in late crystallisation products of alkali magmas especially syenites and nepheline syenites.

Physical Properties of Pyroxenes

- Crystal systems: Orthorhombic and Monoclinic
- Form/Habit: Stout prisms common, showing {100}, {010} and {110} forms terminating with faces of the form {101}.

Colour:

For the orthopyroxenes; the Mg-rich varieties are grey, green, brown, yellow or colourless, while Fe-rich varieties are brownish green or brownish black.

- For the clinopyroxenes; Diopside is white, greenish or dark green. Augite is colourless to pale brown.
- Cleavage: Two good prismatic cleavages {110} good meeting at about 90° in a basal section.
- Fracture: Uneven.
- Lustre: Vitreous to submetallic.
- Hardness: From 5.0 – 6.0
- S.G: Ranges from 3 – 3.96

b. Pyroxenoid group

The pyroxenoid minerals are not structurally related to the pyroxenes. They are chain silicates with more complex structures. Examples include Wollastonite, Pectolite and Rhodonite.

Wollastonite
Composition; CaSiO3
Physical Properties

- Crystal system: Triclinic
- Form/Habit: Columnar or fibrous crystals elongated along the b axis.
- Colour: White, grey, yellowish brown or brownish red.
- Cleavage: {100} perfect; {001} and {102} good.
- Lustre: Vitreous but pearly on cleavage planes.
- Hardness: 4.5 - 5.0
- S.G: 2.87 – 3.09

ii. Double chain silicates (Amphibioles)

- The amphiboles are a group of hydroxylated double chain inosilicates in which some of the OH-groups are substituted by chlorine and Fluorine.
- There are both Orthorhombic and monoclinic members in this group.
- The double-chain structure allows a great level of elemental substitutions.
- The general structural composition is $[Si_4O_{11}]$ with some Si being replaced by Al^{3+}.
- The chains are stacked parallel to the c crystallographic axis and joined together by ions (called A, X and Y types) occupying particular lattice sites. The A, X and Y ions in Amphiboles are made up of; Mg^{2+}, Fe^{2+}, Ca^{2+}, Al^{3+}, Fe^{3+}, Mn^{4+}, Ti^{4+}, K^+ and sometimes Na^+.
- There are 3 main types of amphiboles;
  - Ca-poor amphiboles
  - Ca-rich amphiboles
  - Alkali amphiboles

- All amphiboles are prismatic and have 2 perfect prismatic cleavages which meet at 124° in the basal section.
- Their basal section here has 6 sides.

a. The Ca-poor amphiboles
- These are amphiboles in which Calcium and sodium are about equal to zero (Ca + Na = 0).
- They include orthorhombic amphiboles and the Ca-poor monoclinic amphiboles.
- Their general formula is $X_2Y_5Z_8O_{22}(OH,F)_2$, where X = Mg or Fe^{2+}, Y = Mg, Fe^{2+}, Fe^{3+}, Al etc. and Z = Si or Al.

Occurrence of Ca-poor amphiboles
They occur in high grade regionally metamorphosed basic and ultrabasic igneous rocks by Mg and Fe metasomatism of politic sediments. They are found in amphibolites, hornblendes, gneisses and hornfelses often in association with cordierite.

b. The Ca-rich amphiboles
These are amphiboles in which Calcium is greater than sodium (Ca > Na).

They are monoclinic and include tremolite – ferroactinolite and the hornblende group of minerals.

Their general formula is $AX_2Y_5Z_8O_{22}(OH,F)_2$, Where:
- A= Na (often no sodium is present and A=0),
- X = Ca, Y = Mg, Fe, Al etc and
- Z = Si or Al.

Occurrence of Ca-poor amphiboles
- Tremolite and actinolite are metamorphic minerals formed during thermal or regional metamorphism especially involving impure dolomitic limestones. They occur in relatively low grade metamorphic rocks and actinolite is a characteristic mineral of greenschist facies, occurring with the common hornblendes.

Actinolite may also occur in blueschists, along with glaucophane, epidote and albite.
- Hornblendes are primary minerals in intermediate and some acid plutonic igneous rocks.

The Alkali amphiboles
These are amphiboles in which sodium is greater than sodium Calcium (Na > Ca).

Their general formula is $AX_2Y_5Z_8O_{22}(OH,F)_2$, Where:
- A= Na or K,
- X = Na (or Na and Ca),
- Y = Mg, Fe, Al etc and
- Z = Si or Al.

The occurrence of Alkali amphiboles
- Glaucophane is the essential amphibole in blueschists which form under conditions of high pressure and low temperature in metamorphosed sediments at destructive plate margins.
- A rare amphibole, richterite is found in skarns and in thermally metamorphosed limestones.

2.5.7.5 Phylosilicates (Sheet Silicates)
This group of silicates result when three Oxygen atoms are shared between adjacent tetrahedra forming sheets.

a. Mica Group
- The most important members of this group are Muscovite and Biotite.
- The most distinctive property of mica is a perfect basal cleavage which makes micas able to be separated into thin and flexible sheets.
- Their general formula is $X_2Y_{4-6}Z_8O_{20}(OH,F)_4$,

Where:
- X = Na or K,

- Y = Mg, Fe^{2+}, Fe^{3+} or Al and
- Z = Si or Al.

The specific gravity values of the mica group range from 2.7 to 3.3, and the hardness values occur in the range 2.0 – 4.0.

The micas differ from the chlorites and other micaceous minerals in several ways including;

Their alkali content
The elastic properties of cleavage flakes of micas and

Some of their optical properties.

Muscovite is used in electrical and visual applications.

Muscovite
Composition: $K_2Al_4[Si_6Al_2O_{20}](OH,F)_4$;
Al substitutes for Silicon in the Z group.
Physical properties
- *Crystal System*: Monoclinic, pseudohexagonal
- *Form/Habit*: Occurs as six sided hexagonal plates or as disseminated scales or massive.
- *Colour*: Colourless, white or pale yellow.
- *Cleavage*: One perfect cleavage parallel to the basal plane with large and thin laminae being easily separated.
- *Lustre*: Pearly
- *Transparency*: Transparent and translucent when held up to bright light
- *Tenacity*: Thin flakes are both flexible and elastic
- *Hardness*: 2.5 – 3.0
- *S G:* 2.77 – 2.88

Occurrence
- Muscovite occurs as a primary mineral in acid igneous plutonic rocks like granite and pegmatite.
- It is a common constituent in detrital sedimentary rocks, especially the arenites (sandstone).
- In metamorphic rocks, muscovite occurs in low grade metamorphic rocks where they are derived from pyrophylite or illite (clay minerals). They remain in these rocks as the grade increases and is a common constituent of schist and gneisses. At very high temperatures (above 600°C) muscovite becomes unstable, breaking down in the presence of Quartz to give K-feldspar and sillimanite.

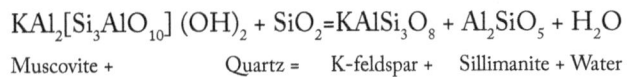

$$KAl_2[Si_3AlO_{10}](OH)_2 + SiO_2 = KAlSi_3O_8 + Al_2SiO_5 + H_2O$$

Muscovite +   Quartz =   K-feldspar +   Sillimanite + Water

Biotite
Composition: $K_2(Mg, Fe^{2+})_{6-4}(Fe^{2+}Al, Ti)_{0-2}[Si_{6-5}Al_{2-3}O_{20}]OH,F)_4$

Biotite is a ferromagnesian mineral because it contains iron (Fe) and magnesium (Mg)
Physical properties
- *Crystal System*: Monoclinic, pseudo hexagonal
- *Form/Habit:* Six sided prismatic crystals common, tabular parallel to (001).
- *Colour*: Black or dark green in thick crystals, but in transmitted light thin laminae appear brown green or blood red.
- *Cleavage:* {001} perfect.
- *Lustre*: Splendent and pearly on the cleavage.
- *Transparency*: Transparent to opaque.
- *Tenacity*: Thin cleavage laminae are flexible and elastic
- *Hardness*: 2.5 – 3.0
- *S G:* 2.7 – 3.30

Occurrence: Biotite is a common mineral in a variety of rocks.
- It forms from chlorite in metamorphosed pelitic rocks and exists over a wide range of regional metamorphic conditions with its magnesium content increasing with increase

in metamorphic grade.

- It is a primary mineral in acid and intermediate plutonic, hypabyssal and extrusive igneous rocks.
- Biotite is a common mineral in clastic sedimentary rocks, particularly arenaceous rocks, but it is prone to oxidation and degradation.

Uses of Micas

- Because at high heat resistant, elastic and platy nature micas are used in electrical equipment as insulators.
- They are also used in lamps, furnace because they are stable at high temperatures.

b. The Chlorite group

Chlorites are phyllosilicate minerals similar to micas in having the fundamental $[Si_4O_{10}]$ sheet but do not contain any alkali. They are considered as hydrous silicates of aluminium (Al), iron (Fe) and magnesium (Mg).

Generally, chlorite minerals crystallise in the monoclinic system and are often pseudohexagonal. Their colour is green and they all poses a perfect {001} basal cleavage which gives cleavage flakes which are flexible but not elastic. They have an average hardness of ≈ 2.

Chlorite

Composition: $(Mg, Al, Fe^{2+})_{12} [(Si,Al)_8 O_{20}](OH)_8$

Physical properties

- **Crystal System**: Monoclinic, pseudohexagonal
- **Form/Habit**: Tabular crystals common. It also occurs as granular masses and disseminated scales in metamorphic rocks.
- **Colour**: Colourless, or green.
- **Cleavage**: Perfect {001} basal cleavage.
- **Lustre**: More or less pearly.
- **Transparency**: Subtransparent to opaque.
- **Tenacity**: Cleavage flakes are flexible and not

elastic
- **Hardness**: 2.0 – 3.0
- **S G:** 2.6 – 3.3

Occurrence

- Chlorite is a common primary mineral in low-grade regional metamorphic rocks such as **greenschists** where they change to biotite with increasing metamorphic grade.
- In igneous rocks, chlorite is mostly a secondary mineral formed from the hydrothermal breakdown of pyroxenes, amphiboles and biolite. It may also occur as an infilling in lava flows and as a primary mineral in some low-temperature veins.
- Chlorites are common in argillaceous sedimentary rocks, where they occur with clay minerals.

c. Clay Minerals

Clay minerals are important products from the weathering of rocks. Feldspars in particular give rise to clays, with K-feldspar reacting in the presence of water to give illite and plagioclase feldspar reacting in a similar manner to give montmorillonite. If excess water is present, both reactions will eventually produce kaolinite as a final product.

The clays can either remain in the place of weathering to form residual clays or they may be transported by various agents (water, wind and ice) and deposited as beds of clay in the sea and lakes or as superficial deposits of boulder clays or as loess or **adobe deposits**[4] in desert.

Clays have the property of absorbing water and becoming either plastic or liquid. The plastic or liquid

---

[4] Adobe deposits: A silty clay, often calcareous, found in dry, desert-lake basins. This fine-grained sediment is usually deposited by desert floods which have eroded wind-blown loess deposits. The term is of Spanish origin.

limits are defined by values called *Atterberg limits*[5] .

Clays also become hard when heated to a suitable temperature. All clay minerals contain $[Si_4O_{10}]$ tetrahedral sheets, similar to the micas and tend to occur as minute flaky crystals.

All clay related minerals have a very small grain size in the natural state making identification by optical techniques virtually impossible and so they are usually studied using X-ray diffraction techniques (SRD) by using a scanning electron microscope (SEM) or an electron microscope.

Common examples of clays include: Kaolinite, Illite, Montmordonite, Vermiculite, Sepiolite (Meerachaum) and Allophane.

Kaolin, Kaolinite or China clay (Locally called "Calabar Chalk")
Composition: $Al_4[Si_4O_{10}](OH)_8$
Physical properties
- *Crystal System*: Triclinic or monoclinic.
- *Form/Habit*: Pseudohexagonal tabular crystals. Usually massive.
- *Colour*: white when pure, grey or yellowish when impure.
- *Cleavage*: Perfect {001} basal cleavage.
- *Lustre*: Dull and Earthy.
- *Feel*: Greasy feel, often very soft material, crumbling to powder when pressed between the fingers.
- *Smell*: Clayey smell
- *Hardness*: 2.0 – 2.5
- *S G:* 2.61 – 2.68

---

[5] Atterberg limits: Series of thresholds which are observed when the water content of a soil is steadily changed. Knowledge of these limits is important for understanding and predicting hillslope failure. The limits were described by the Swedish soil scientist Albert Mauritz Atterberg (1846–1916).

Occurrence:
- Kaolin or China clay consists partly of crystalline and partly of amorphous material.
- It is formed from the alteration of feldspars in granites. This alteration may be caused either by the weathering process or by pneumatolytic action of gases on feldspars.

Uses of clays
- Clays are used for the manufacture of fine porcelain and china (for example, breakable plates, bowls and tea cups), porcelain fittings
- They are also used as fillers in paper, rubber and paint manufacture.
- A variety of clays known as sepiolite (meerachaum) was formerly used in North Africa as a substitute for soap.

2.5.7.6 Tectosilicates (framework silicates)
There are 5 groups of tectosilicates as follows:
1. The silica group (Quartz)
2. The feldspar group
3. The feldsparthoid group
4. The zeolite group and
5. The scapolite group.

The silica group (SiO₂):
This group has a structure in which the Silicon – Oxygen (Si-O) tetrahedra build up a 3 dimensional framework, with each tetrahedron sharing all their 4 Oxygen atoms with neighbouring tetrahedra. Silica exists in many forms including the following:
- Crystalline silica group for example; Quartz (alpha, α and beta, β), tridymite, crystobalite, coesite and stishovite.
- Cryptocrystalline group for example chalcedony, jasper, flint, opal.

i. Crystalline silica group
There are three main forms of crystalline silica

that exists as polymorphs which are Quartz, tridymite and cristobalite. These polymorphs are stable at different temperatures and pressures and can be represented by a pressure/temperature diagram which shows their stability ranges and relationships.

Quartz exists in 2 forms; Alpha (α) Quartz or Low Quartz, and Beta (β) Quartz or High Quartz.

The phase relationship shows that:
- Alpha (α) Quartz which is the lowest temperature form of Quartz is stable at temperatures below 573 oC and inverts to Beta (β) Quartz at 573° C.
- Beta (β) Quartz is stable between 573° C to 867°C and inverts to tridymite at 870°C.
- Tridymite is stable between 867°C to 1470oC and above this temperature, it inverts to cristobalite.
- Cristobalite is stable between 1470°C to 1713°C and above this temperature, it melts and the Quartz liquidus boundary is reached. This is demonstrated in the diagram below;

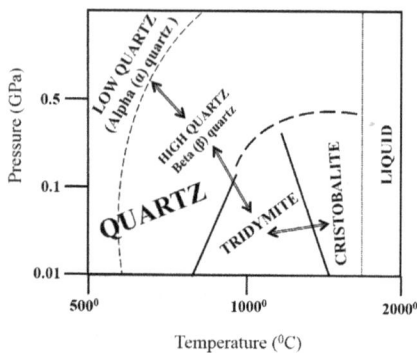

Figure 2.22: The pressure/temperature (P-T) diagram for Crystalline silica polymorphs (Quartz, Tridymite and Cristobalite)

Pressure is measured in GPa[6] .

The other members of the crystalline group of silica are coesite and stishovite which form polymorphs at very high pressures. Both of them are recorded from meteor impacts and rarely occur in normal terrestrial rocks.

The polymorphs of Quartz act as geothermometers since they give indication of pressure and temperature conditions at which the rocks bearing them were formed.

Quartz
Composition; SiO_2
Physical Properties
- *Crystal system*: Trigonal
- *Form/Habit*: Usually as hexagonal prisms terminating with rhombohedral faces which resemble hexagonal bipyramids.
- *Colour*: Colourless or white when pure but in the presence of impurities, it shows a wide range of colours giving rise to different varieties. These impurities are due to inclusions of other minerals or liquids resulting to Amethyst (violet or purple), rose Quartz (pink), Citrine (light brown).
- *Cleavage*: None.
- *Fracture*: Conchoidal
- *Lustre*: Vitreous, occasionally resinous.
- Transparency: Transparent when clear, to translucent when white or coloured.
- *Hardness*: 7 (one of the minerals on Mohs' Hardness scale)
- *S.G*: 2.65

Diagnostic characteristics include the presence of a well-developed conchoidal fracture, the absence of cleavages and an external average hardness of 7 on Mohs's

---

[6] GPa = Gigapascals. 1GPa = 10 kbar

scale of hardness.

Occurrence:

- Quartz is an essential constituent of acid igneous plutonic rocks such as granites, granodiorites and pegmatites. It may sometimes appear as shapeless interstitial grains in some diorites and gabbros.
- It occurs as phenocrysts with corroded edges in extrusive and hypabyssal rocks such as rhyolites, dacites, pitchstones and various porphyrites.
- It is a common gangue mineral in hydrothermal and other veins, accompanying the economic ore mineral.
- Quartz is a common detrital mineral because of its hardness, lack of cleavage and stability.
- It is an essential mineral in the coarser type of terrigenous rocks such as conglomerates, arenites etc, and also occurs in fine grained varieties such as siltstone and mudstones, although its identification may be difficult.
- Authigenic Quartz may form during sedimentary diagenesis, often growing around pre-existing Quartz grains.
- Quartz occurs in many metamorphic rocks, especially pelites and psammites and remain until the very highest grades when it enters into the reaction.

Uses of Quartz and its polymorphs

- * It is used in electronic industries as oscillators for frequency control. This is because of its piezoelectric properties.
- * It is used as abrasives on sandpaper due to its hardness.
- * It is used for jewellery and also for making screens (glass) for deep sea vessels.
- * Optically clear Quartz is used in the manufacture of lenses and prisms.
- * The coloured variety such as opal is cut and used as a precious gem.

ii. Cryptocrystalline silica group
Composition: $SiO_{2-n}H_2O$

Most varieties are mixtures of cryptocrystalline silica and hydrous silica; from SiO_2 (Chalcedonic silica) to opal (Hydrous silica).

Physical Properties

- *Crystal systems*: none
- *Form/Habit*: the different varieties may occur either as fibrous structures or veinstones or as nodules in sedimentary or extrusive rocks.
- *Colour*: Variable, displaying a play of colours.
- *Cleavage*: None.
- *Fracture*: Conchoidal
- *Lustre*: Most are waxy, some are subvitreous and opal is opalescent (iridescent).
- *Transparency*: Transparent, translucent or opaque.
- *Hardness: 5.5 – 7*
- *S.G:* Ranges from 1.99 – 2.67

b. Feldspar group of Tectosilicates

Feldspars constitute the most important group of tectosilicates and rock forming minerals occurring in igneous, sedimentary and metamorphic rocks. They are formed by the substitution of Silicon by aluminium in the SiO_2 structure of the tectosilicates. When some Silicon ions are substituted for AL^{3+} a charge deficit is created which is balanced by the introduction of cations like Na^+, K^+ and Ca^{2+} depending on the level of substitution as follows;

If the substitution is low about 25%, univalent cations like K^+ and Na^+ will be introduced and the feldspar formed are known as *Alkali feldspars*.

$$Si_4O_8 (Al_2Si_2O_8)^{-1} + K^+ \text{ or } Na^+KAl Si_3O_8 \text{ or } NaA_1 Si_3O_8$$

On the other hand, if the substitution is about

50% divalent cations are introduced because a net charge of -2 is created for example Ca_{2+}.

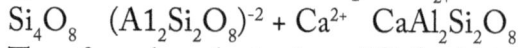

$$Si_4O_8 \quad (Al_2Si_2O_8)^{-2} + Ca^{2+} \quad CaAl_2Si_2O_8$$

Therefore, the substitution of Si for Al in tectosilicates brings about the following types of feldspars:

- Potash Feldspar ($KAlSi_3O_8$)
- Sodic feldspar ($NaAlSi_3O_8$)
- Calcic feldspar ($CaAl_2Si_2O_8$)

These three groups of feldspar can be replaced in two main groups as follows

- *Alkali feldspars* which contain Na+ and K+ and occupy a range of composition between sodic feldspars (albite), $NaAlSi_3O_8$ and potash feldspars (Orthoclase), $KAlSi_3O_8$.
- *Plagioclase feldspar* which occupy a range of compositions between Sodic and calcic feldspar (anorthite), $CaAl_2Si_2O_8$.

i. Plagioclase feldspars

Plagioclase feldspar contains calcium and sodium ions (Na+ and Ca^{2+}). They are formed by a solid solution which exists between Anorthite ($CaAl_2Si_2O_8$) and Albite ($NaAl_3Si_3O_8$) because Na_+ and Ca^{2+} easily substitute each other at all temperatures. This substitution brings about the formation of various intermediate minerals between anorthite and albite called plagioclases. It is rare to have pure anorthite or pure albite. However, pure anorthite has 100% Ca and 0% Na, while pure albite has 100% Na and 0% Ca anorthite. So a complete range of plagioclase feldspars can be described with albite at one end and anorthite at the other end.

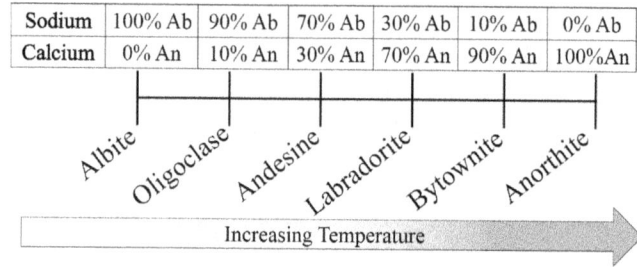

Sodium	100% Ab	90% Ab	70% Ab	30% Ab	10% Ab	0% Ab
Calcium	0% An	10% An	30% An	70% An	90% An	100%An

Albite Oligoclase Andesine Labradorite Bytownite Anorthite

Increasing Temperature

Figure 2.23: The Plagioclase isomorphic series.

Anorthite is a high temperature variety of plagioclase while albite is a low temperature. So during crystallisation of plagioclase with falling temperature, a solid solution series is formed in which there exist early formed crystals of anorthite (calcium rich plagioclase) in suspension in the melt of albite (Na rich).

Within this solid solution Na+ are substituting for Ca^{2+} in the early form crystals. This can be seen in the phase diagram for the plagioclases below.

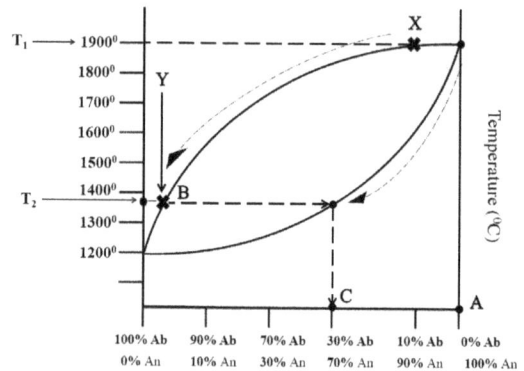

Figure 2.24: Phase diagram for anorthite – albite isomorphic series.

From the above phase diagram;

- Crystallisation begins at point X, 1900°C (T_1). The composition of the crystals at point X are known at point A which is 100% Anorthite (An_{100}) and 0% Albite (Ab_{00}); that is, the plagioclase feldspars produced are richer in

Calcium (Ca) with no sodium (Na).

- As the melt progressively cools, its calcium content also gradually decreases while its sodium content increases that is why at the portion at Y the crystals B have a composition C of 70% Anorthite and 30% Albite.

Therefore, during the cooling process of a plagioclase melt, calcium rich members crystallise first, while sodium rich members crystallise last.

ii. Alkali feldspars

The alkali feldspar contain Na^+ and K^+ include a group of minerals between Albite ($NaAlSi_3O_8$) and orthoclase ($KAlSi_3O_8$)

At high temperature there exist a solid solution between albite and orthoclase where K and Na are substituted for one another. This substitution forms a variety of alkali feldspar including; Anorthoclase, Microcline, Sanidine Orthoclase.

However, at low temperature there is no substitution between Na and K. Microcline is a low temperature K-feldspar while sanidine is a high temperature variety. Feldspars are usually represented by a triangular phase diagram whose apexes are occupied by Orthoclase, ($KAlSi_3O_8$), Albite ($NaAlSi_3O_8$) and Anorthite ($CaAl_2Si_2O_8$). These ternary diagrams show the possibilities of ionic substitution between the feldspars.

showing solid solution possibilities under high temperature conditions

At high temperatures there exist a 2 solid solutions;

Between the alkali feldspars (Albite and Orthoclase) where K and Na are substituted for one another and

Between the plagioclase feldspars (Anorthite and Albite) where Ca and Na are substituted for one another.

At low temperatures however, there exist only a single solid solution between plagioclase feldspars (Anorthite and Albite).

The alkali feldspars are immiscible at low temperatures and so no solid solution exists between them.

Anorthite and Orthoclase do not form solid solutions at all temperatures due to several factors including their difference in ionic size.

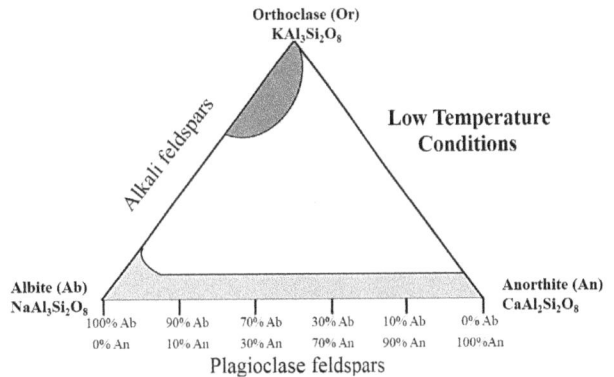

Figure 2.26: Ternary diagram for the feldspars showing solid solution possibilities under low temperature conditions

Perthites and Antiperthites

As already mentioned above, Na and K are not mixable at low temperature but at high temperature they become mixable. Most alkali feldspars are usually formed with inclusions in them called immiscibility structures or ex-solution structures. These

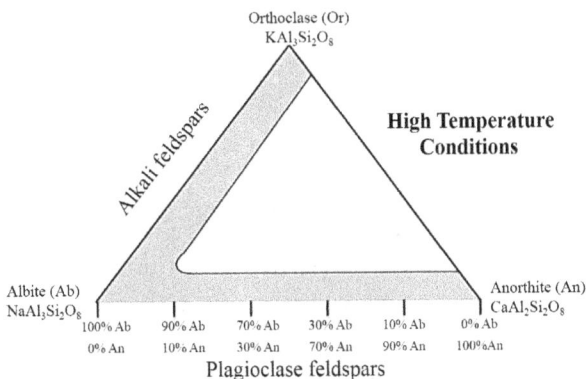

Figure 2.25: Ternary diagram for the feldspars

structures are pethites and antiperthites.

Perthites: Perthites are intergrowths of sodic plagiocaaes (albite) in a matrix of potash feldspar (Orthoclase). The sodic plagioclase forms leticels or laths in a host of potash feldspars because during cooling, as temperature begins to drop, the originally homogenous alkaline feldspar melt becomes immiscible. The Na-rich plagioclase phase segregates from the host potash feldspar.

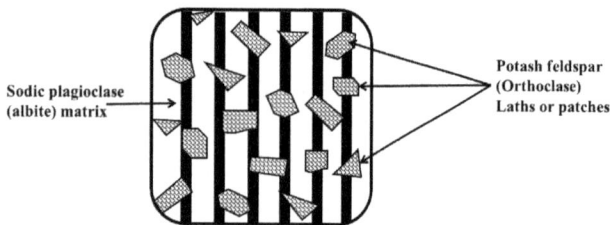

Figure 2.27: Perthites.

Antiperthites: These are inter-growths of orthoclase or potash feldspar in a host of plagioclase.

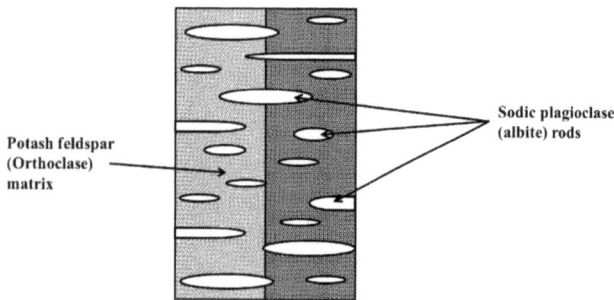

Figure 2.28: Anti-perthites.

The origin of perthites and antiperthites can be suggested as follows

Feldspars which were homogenous at high temperature becomes cons at low temperature and one component began to crystalise out

By replacement where the early formed potash feldspar react with the soda rich liquid.

Physical properties all feldspars

Crystal systems: Most K-feldspars are Monoclinic, all plagioclases are triclinic.

Form/Habit: Usually prismatic (although those crystallized at high temperatures may have a tabular habit).

Colour: Variable, usually white or light coloured, but orthoclase is often pink or reddish, and Ca-rich plagioclase is often dark grey.

Cleavage: all feldspars possess two cleavages, a prismatic cleavage parallel to {010} and a basal cleavage parallel to {001} which meet at ~ 90° on the {100} face. Several partings may also occur.

- Fracture: Conchoidal or uneven and splintery.
- Lustre: Vitreous to pearly on cleavage planes and sometimes dull.
- Transparency: subtransparent to translucent and opaque. The higher temperature members are more transparent than the lower temperature members.
- Hardness: 6.0 – 6.5
- S.G: Alkali feldspars = 2. 56, plagioclase feldspars = 2. 63

Diagnostic characteristics include the presence of a well developed conchoidal fracture, the absence of cleavages and an external average hardness of 7 on Mohs's scale of hardness.

Occurrence

The range of composition in the feldspars has led them to be used as a means of classifying igneous rocks, since they are absent only from certain ultramafic and ultra-alkaline igneous rock types and Carbon ates.

In metamorphic rocks, feldspars are absent only from some low grade pelites, pure marbles, pure Quartzites and most eclogites.

Feldspars are common in arenaceous sedimentary rocks, but less common in argillaceous types.

c) Feldsparthoid group of tectosilicates

Feldsparthoids are a group of tectosilicates similar to feldspars in structure and chemical composition but containing less amount of silica (SiO_2). For this reason, they are said to be unsaturated with silica.

Their occurrence is restricted to undersaturated alkali igneous rocks, except for some (like Lazurite) which occurs in contact metamorphosed limestone.

2.6 Non-Silicate Minerals

Non silicate minerals are minerals that lack the silicate structure (SiO_4) which is the fundamental structure of the silicates. The non-silicate minerals are not as abundant as the silicates but they however appear to be more economically important. That is, most of the non-silicate minerals exist as ores and can be mined and used for economic benefits.

Non-silicate minerals are formed by the combination of certain anions or radicals with metals or groups of metal. For the purpose of classification, non-silicate minerals are treated under the anion or radical that constitute them. Non-silicate minerals include;

- Sulphides
- Carbon ates
- Oxides
- Sulphates
- Halides
- Phosphates

There are also some non-silicate minerals which do not exist as compounds for example Cu, Gold, and Silver etc. Such minerals are native elements (native minerals)

2.6.1 Sulphide Minerals

This is a group of minerals in which the element sulphur (S) is in combination with one or more metallic elements. Simple sulphides include the common ore minerals; galena (PbS), sphalerite (ZnS), and pyrite (FeS_2). Two metallic cations may also be present, as in chalcopyrite ($CuFeS_2$). The most important sulphide minerals are:

- Galena, PbS
- Sphalerite (ZnS)
- Pyrite, FeS_2
- Chalcopyrite, $CuFeS_2$

Galena, lead, blue lead
Composition – PbS
Physical properties

- Crystal systems: Cubic
- Form/Habit: Cubes are common often modified by other forms in the cubic system like octahedrons. Also massive or finely granular.
- Colour: lead grey
- Streak: lead grey
- Cleavage: Perfect {100} cubic cleavage.
- Fracture: Flat, even or subconchoidal.
- Lustre: Metallic on fresh surfaces, but dull when exposed due to tarnish.
- Hardness: 2.5
- S.G: 7.4 – 7.6

Occurrence

- Galena often occurs in association with sphalerite (ZnS) in minerals in hydrothermal conditions.
- They also occur as replacement minerals in different rock types.
- Galena of sedimentary origin are not important.

Uses

- Galena is the most important ore of lead.
- Sphalerite, Zinc blende, Black Jack
- Composition: ZnS
- Physical Properties
- Crystal systems: Cubic
- Form/Habit: Tetrahedra or dodecahedra commonly massive and compact sometimes

botryoidal or fibrous.
- Colour: black or brown sometimes yellow or white.
- Streak: White or reddish brown
- Cleavage: Perfect {110}.
- Fracture: Conchoidal.
- Lustre: Resinous to adamantine.
- Transparency: Transparent to translucent or opaque
- Tenacity: Brittle
- Hardness: 3.5 – 4.0
- S.G: 3.9 – 4.2

Occurrence
- Sphalerite is common in stratabound veins and massive sulphide deposits where it is associated with galena.
- It is found in association with other sulphides like pyrite, Galena, chalcopyrite in calcareous nodules or veinlets.
- Sphalerite is the most important ore of Zinc.

Pyrite (Iron pyrites)
Composition: FeS_2
Physical Properties
- Crystal systems: Cubic
- Form/Habit: Cubic and pentagonal dodecahedron forms occur in crystals. Faces are usually striated with the striations on one face being perpendicular to those of adjacent faces.
- Colour: Bronze yellow to pale brass yellow.
- Streak: Brownish black or greenish black
- Cleavage: None.
- Fracture: Conchoidal or uneven.
- Lustre: Metallic or splendent.
- Transparency: Opaque
- Tenacity: Brittle. Sparks are produced when pyrite is hit with steel.
- Hardness: 6.0 – 6.5
- S.G: 4.8 – 5.1

Occurrence
- Pyrite is a frequent constituent of many ore veins.
- Occurs in contact metamorphic rocks as a result of pyrometasomatism
- Also occurs in large intrusive igneous rocks as a result of magnetic segregation.
- Pyrite deposits of sedimentary origin are represented by oolithic pyrites.

Chalcopyrite
Composition: $CuFeS_2$
Physical Properties
- Crystal systems: Tetragonal
- Form/Habit: Twin crystal resembling tetrahedral usually massive in habit
- Colour: Brass Yellow
- Streak: Greenish black
- Lustre: Metallic
- Fracture: Conchoidal or uneven
- Cleavage: Poor {011} and {111}
- Hardness: 3.5 – 4.0
- S.G: 4.1 – 4.3

Occurrence
- Chalcopyrite is the principal commercial source of copper.
- It occurs as hydrothermal and metasomatic veins.
- It occurs with other sulphides and skarn minerals at or near the contact between bodies of intrusive granodioritic rocks and limestone.

2.6.2 Oxide Minerals
Oxide minerals are a group of minerals in which Oxygen is combined with one or more metals to give simple and multiple oxides respectively.

The most important oxides are haematite, magnetite, corundum, limonite and Ilmenite

i. Simple oxides

These are oxides in which Oxygen combines with only one type of metal. They include the following;

Corundum

Composition: Al_2O_3

Physical Properties Crystal systems: Trigonal

- Form/Habit: It occurs as barrel-shaped or pyramidal crystals and steep hexagonal crystals. Also it can be massive and granular
- Colour: Shows varied colours; grey, greenish or red. Precious varieties ruby and sapphire are red and blue respectively.
- Lustre: Vitreous, sometimes faces are dull
- Fracture: Conchoidal or uneven
- Cleavage: none
- Transparency: Transparent to subtransparent
- Hardness: 9
- S.G: 3.98 – 4.02

Occurrence

- Corundum occurs in metamorphosed bauxite deposits.
- In igneous rocks, it occurs in silica poor rocks like syenites, and other undersaturated alkali igneous rocks.
- In sedimentary rocks, gem varieties and corundum exist as rounded pebbles in alluvial deposits.

Uses

- Corundum, apart from diamond, is the hardest known mineral and is used as an abrasive.
- Grinding wheels are made using crushed corundum.
- Artificial corundum is made by fusing bauxite in an electric furnace.
- Coloured varieties are of corundum are important gemstones.

Haematite (Kidney Ore)

Composition: Fe_2O_3

Physical Properties

- Crystal systems: Trigonal
- Form/Habit: Rhombodecahedron forms often massive, tabular, foliaceous
- Colour: Steel grey to iron black. In transmitted light thin platy crystals appear blood red.
- Streak: Cherry red to reddish brown.
- Lustre: Specular iron ore are highly splendent. Amorphous varieties are dull and Earthy and fibrous varieties are silky.
- Fracture: Subconchoidal to uneven
- Cleavage: Poor parallel to $\{10\bar{1}1\}$ and $\{0001\}$
- Hardness: 5.5 – 6.5
- S.G: 4.9 – 5.3

Occurrence

- Haematite occurs in igneous rocks as accessory minerals especially in basic and ultrabasic rocks.
- It also occurs as hydrothermal veins
- Haematite widely occurs in sedimentary rocks as current minerals e.g. in sandstone
- Haematite occurs in sedimentary rocks as banded iron stone
- In igneous and metamorphic rocks, it occurs in association with titanium oxide
- Haematite occurs in pockets and hollows, replacing limestone.
- The greatest haematite deposits in the world in the Lake superior region were formed from the alteration and concentration of iron silicates and iron Carbon ates of sedimentary origin

Distinction – Red streak and hardness 5.5 – 6.5

Ilmenite
Composition: $FeTiO_3$
Physical Properties
- Crystal systems: Trigonal
- Form/Habit: Thin plates or scales, or massive; sometimes as heavy residue grains in sand.
- Colour: Iron black.
- Streak: Black to brownish black.
- Lustre: Submetallic
- Fracture: Conchoidal
- Cleavage: Parting parallel to {111}
- Transparency: Opaque
- Hardness: 5.0 – 6.0
- S.G: 4.5 – 5.0

Occurrence
- Ilmenite occurs as an accessory mineral in basic igneous rocks like gabbro and diorite.
- It also occurs as dyke-like bodies or veins in association with magnetite and chalcopyrite
- Many important deposits of Ilmenite occur as detrital placer deposits in sedimentary rocks
- Magnesium-rich Ilmenites occur in kimberlites and also in contact metamorphosed rocks.

Distinction: from magnetite by its non-magnetic and from haematite by colour and streak.

ii. Multiple oxides:
These are oxides in which Oxygen combines with more than one type of metal. They include the following;

Limonite (brown haematite)
Composition – $FeO(OH).nH_2O$
Physical Properties
- Crystal systems: None (amorphous)
- Form/Habit: Mammilated[7] or stalactitic habits, with radiating fibrous structure resembling that of haematite.
- Colour: Black or brownish yellow when Earthy.
- Streak: Yellowish or yellowish brown.
- Lustre: Submetallic to silky but usually dull
- Hardness: 4.5 – 5.5
- S.G: 2.7 – 4.3

Occurrence
It occurs widely as weathered products other iron minerals therefore it is frequently mixed with many substances like magnetite and manganese oxide.

Magnetite
Composition: Fe_3O_4.
It usually contains Ti and Mg with minor amounts of Mn, Ca and Ni.
Physical Properties
- Crystal systems: Cubic
- Form/Habit: Octahedral also granular and massive
- Colour: Black.
- Streak: Black.
- Lustre: Metallic to submetallic
- Fracture: Subconchoidal to uneven
- Cleavage: Parting parallel to {111}
- Transparency: Opaque
- Tenacity: Brittle.
- Magnetism: Strongly magnetic
- Hardness: 5.5 – 6.5
- G: 5.2

Occurrence
- Magnetite occurs as a primary constituent of

---

[7] Mammillary (mammillated): Applied to the physical habit of a mineral which has grown from radiating crystals to give curved or rounded surfaces.

most igneous rocks.

- Most are considered to be the result of magmatic segregation. Magnetite occurs in metamorphic rocks derived from ferruginous sediments and is associated with Quartz and chlorite.
- Magnetite may be formed by metasomatic replacement of limestone and in contact metamorphic rocks.
- Magnetite is also a constituent of many veins and is found in residual clays and in placer deposits such as the "black-sands" formed by the degradation of earlier deposits.

Uses
It is one of the most valuable ore of iron

2.6.3 Hydroxides
These are minerals whose chemical composition includes the OH radical and often water molecules, whose presence gives hydrated oxides. They include the following;

Brucite
Composition: $Mg(OH)_2$
Physical Properties
- Crystal systems: None (amorphous)
- Form/Habit: Crystals when present, have a prismatic and broad tabular habit.
- Colour: White, often bluish, greyish and greenish.
- Lustre: Pearly
- Transparency: Translucent to subtranslucent
- Hardness: 2.5
- S.G: 2.9

Occurrence
Brucite is found in metamorphosed contact impure limestones called pencatites.
Uses:

Brucite is used for the production of magnesium and refractories[8].

Gibbsite (Hydrargillite)
Composition: $Al(OH)_3$
Physical Properties
- Crystal systems: Monoclinic
- Form/Habit: Crystals are usually occurring as concretions.
- Colour: White.
- Hardness: 3
- S.G: 2.35

Occurrence
- Gibbsite occurs in deposits of bauxite and as an alteration product from aluminosilicates.

Bauxite
- Composition: It is a mixture of aluminium hydroxides and gibbsite in different amounts, together with impurities of iron oxide, phosphorus compounds and TiO_2.
- Physical Properties
- Form/Habit: Amorphous in Earthy granular.
- Colour: Dirty white, greyish, brown, yellow, reddish brown.

Occurrence
- Bauxite results from the decay and weathering of aluminous rocks under typical conditions. It may form residual deposits replacing the original rock or it may be transported from its place of origin and form deposits elsewhere.

Uses
- The principal use of bauxite is for the

---

[8] Refractories are substances that are resistant to depreciation especially decomposition by heat. Refractory minerals are minerals resistant to decomposition by heat, pressure, or chemical attack. Most commonly applied to heat resistance.

manufacture of aluminium.

- Considerable quantities of bauxite are used as abrasives.
- Low grades are used as refractories, as refractory bricks and for furnace and converter linings.

2.6.4 Sulphate Minerals

Group of non-silicate minerals in which the SO_4^{2-} radical is in combination with a number of metal cations. Examples include anhydrous sulphates like barite ($BaSO_4$), and anhydrite ($CaSO_4$) and hydrated sulphates like Gypsum ($CaSO_4.2H_2O$) which is the most common of a number of hydrated sulphates.

i. Anhydrous Sulphates

Barite or Baryte ($BaSO_4$)
Physical Properties

- Crystal systems: Orthorhombic
- Form/Habit: Tabular, massive, lamellar, granular and compact.
- Colour: Colourless or white when pure but often tinged with red, yellow or brown when impure.
- Streak: White.
- Lustre: Vitreous to resinous; sometimes pearly.
- Fracture: Uneven
- Cleavage: 3 perfect cleavage {001}, {210} and {010}
- Transparency: Transparent to Opaque
- Tenacity: Brittle.
- Hardness: 3.0 – 3.5
- S.G: 4.5 (when pure)

Occurrence

- Occurs as a very common gangue minerals in metalliferous hydrothermal veins in association with Quartz, Calcite, fluorite, sphalerite

etc.

- It also occurs as residual nodules resulting from the decay of limestones containing barite veins.

Uses

Barites are used in the manufacture of white paint especially to give white paper.

Anhydrite
Composition: ($CaSO_4$)
Physical Properties

- Crystal systems: Orthorhombic
- Form/Habit: Tabular, massive, lamellar, granular and compact.
- Colour: Colourless or white when pure but often red, grey or bluish tints.
- Lustre: Vitreous sometimes pearly on cleavage plane.
- Fracture: Uneven
- Cleavage: 3 mutually perpendicular cleavages on {010} and {100} perfect and {001} good.
- Transparency: Transparent to subtransparent.
- Hardness: 3.0 – 3.5
- S.G: 2.93 – 3.0

Occurrence

- Anhydrite occurs as a saline residue associated with gypsum and halite.
- Many anhydrite beds form by the dehydration of gypsum.
- Anhydrite is associated with gypsum and sulphur in the 'cap rock' of overlying slat domes.

Uses

Anhydrite is of importance as a fertilizer, in the manufacture of plasters and cements, sulphates and sulphuric acid.

ii. Hydrated Sulphates

Gypsum

Composition: $(CaSO_4.2H_2O)$

Physical Properties

- Crystal systems: Monoclinic
- Form/Habit: Prisms common, flattened parallel to (010). Crystals are combinations of {010}, {110} and {011}. It also occurs in laminated, granular and compact masses and on fibrous habits.
- Colour: Colourless, massive varieties occasionally grey or yellowish
- Lustre: Vitreous to resinous; sometimes pearly.
- Cleavage: Perfect cleavage {010}, imperfect {100} and {ī11}
- Transparency: Transparent to translucent and even opaque
- Tenacity: Sectile.
- Hardness: 1.5 -2.0 can be stretched by a finger nail
- S.G: 2.31

Occurrence

- Gypsum is formed in the following ways;
- As an evaporate deposits in saline environments like lakes, lagoons, etc. Here it forms thick stratified sedimentary beds. In such basins it is usually the first mineral to precipitate in the normal evaporate sequence.
- Gypsum may form near fumeroles and volcanic vents and it may occur in the gossan or oxide zones of metalliferous sulphide deposits if Carbon ate rocks are present.
- Another way in which gypsum is formed is by the hydration of anhydrite.

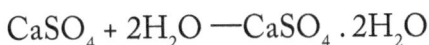

$$CaSO_4 + 2H_2O \longrightarrow CaSO_4 . 2H_2O$$

Uses

Used in the manufacture of plaster boards and other plaster products.

2.6.5 Carbonate Minerals

This is a group of minerals found mostly in limestones and dolomites. Calcite $(CaCO_3)$ is the most abundant and most important. Aragonite has the same formula as calcite but is less stable and shells composed of aragonite change to calcite through geological time. Dolomite(or pearl-spar) is the magnesium-bearing Carbonate commonly found as a rock-forming mineral, $CaMg(CO_3)_2$.

The term 'Carbonate' is frequently used with reference to those sedimentary rocks with 95% or more of either calcite or dolomite, and is synonymous with limestone.

Calcite

Composition: $CaCO_3$

Physical Properties

- Crystal systems: Trigonal
- Form/Habit: Well-formed calcite crystal exhibits 3 main habits;
- Rhombohedron (10ī1)
- Dog-tooth spar, a combination of scalenohedron {213̄1} and prism {10ī0}
- Nail head spar, a combination of flat Rhombohedron (10ī2) and prism {10ī0}
- Also anhedral calcite crystals occur as shapeless grains, fibrous, lamellar and massive and in stalagmitic and stalactitic habits.
- Colour: Colourless or white, sometimes with grey, yellow, blue, red, brown, or black tints.
- Cleavage: Perfect rhombohedral cleavage (10ī1)
- Fracture: Conchoidal but rarely seen because of perfect cleavages
- Lustre: Vitreous to dull.
- Transparency: Transparent to opaque

- Hardness: 3 (a mineral on the Mohs' scale of hardness)
- S.G: 2.71 – 2.94

Occurrence:

- Calcite is one of the most common and most widespread minerals on or near the Earth's surface, where it is the only stable form of CaCO3.
- It is a principal constituent of sedimentary limestones and occurs in Carbon ate shells as fine precipitates and as clastic material.
- In hydrothermal veins, calcite is a common gangue mineral, together with fluorite, Quartz or barite found in association with sulphide ore minerals such as sphalerite and galena.
- Calcite can occur as a primary mineral in some alkali igneous rocks and Carbonates and is a common secondary mineral in basic igneous rocks after the alteration of ferromagnesian minerals by late-stage hydrothermal solutions carrying CO_2.

Uses:

- Calcite finds many different uses according to its purity and character.
- The varieties containing some clayey matter are burnt for cement.
- The purer variety provide lime which is used in many industrial processes such as the manufacture of bleaching powder, calcium carbide, glass, soap, paper and paints.
- Marbles and crystalline limestone and other resistant calcareous rocks generally are important building and ornamental stones.
- Calcium Carbonate is used as a flux in smelting.
- Certain varieties of limestone are used in printing processes.
- Crushed limestones have major use as

agricultural lime used to neutralise the natural acids in the soils.

Dolomite
Composition: CaMg(CO$_3$)$_2$
Physical Properties

- Crystal systems: Trigonal
- Form/Habit: Rhombohedron common, usually with curved surfaces.
- Colour: White, yellow, brown, sometimes red, green, or black.
- Cleavage: Perfect rhombohedral cleavage $(10\bar{1}1)$
- Fracture: Conchoidal or uneven
- Lustre: Crystals are vitreous to pearly, while massive varieties are dull.
- Transparency: Translucent to opaque
- Tenacity: Brittle
- Hardness: 3.5 – 4
- S.G: 2.86 – 3.1

Occurrence

- Dolomite occurs in extensive beds at many geological horizons.
- Dolomites may be deposited directly from sea water, but most Dolomitebeds have been formed by the alteration of limestone where Dolomitereplaces calcite in the process of dolomitisation.
- Under metamorphism, Dolomitemay recrystallize to give Dolomitemarble, but at high temperatures it may decompose to calcite and periclase or brucite, or if silica is present, it may combine with the magnesia to form magnesian silicates such as forsterite, diopside, tremolite etc.
- Dolomiteoccurs in hydrothermal veins with calcite, siderite, fluorite, etc and with ore minerals like sphalerite, chalcopyrite and galena.
- It may occur as secondary minerals in

hydrothermally altered ultramafic igneous rocks.

Uses
- Dolomite is an important building material.
- It is also used for making refractory furnace linings.
- It is a source of Carbon dioxide.

Malachite
Composition: $Cu_2CO_3(OH)_2$
Physical Properties
- Crystal systems: Monoclinic
- Form/Habit: Prismatic crystals are common, sometimes massive.
- Colour: Bright green, shades of green, often concentrically banded.
- Streak: Green of massive varieties, paler than colour.
- Cleavage: Perfect on $\{2\bar{0}1\}$; good on $\{010\}$
- Lustre: silky on fibrous surfaces. Massive varieties are dull. Crystals have adamantine to vitreous lustre.
- Transparency: Crystal varieties are translucent to subtranslucent while massive varieties are opaque
- Hardness: 3.5 – 4
- S.G: 3.9 – 4.1

Occurrence
- Malachite is a secondary mineral found in the zone of weathering or oxidation of copper deposits, lodes or other types, where it is closely associated with azurite.

Uses
- Malachite is a valuable ore of copper.
- It is also cut, polished and used for ornamental purposes.

Azurite
Composition: $Cu_2(CO_3)_2(OH)_2$
Physical Properties
- Crystal systems: Monoclinic
- Form/Habit: Crystals are modified prisms or tabular, parallel to the basal plane, but it is usually massive.
- Colour: Deep azure blue, hence the name azurite.
- Streak: Blue, lighter than the colour.
- Cleavage: Perfect on $\{001\}$; distinction on $\{100\}$ and poor on $\{110\}$
- Lustre: Crystals have adamantine to vitreous lustre, silky or resinous on radiating surfaces. Massive varieties are dull.
- Transparency: Crystal varieties are Transparent while massive varieties are opaque
- Tenacity: Brittle
- Hardness: 3.5 – 4
- S.G: 3.77 – 3

Occurrence
- Azurite is common in the oxide or weathered zone of copper deposits.
- It is always found with cuprite and in association with malachite.

2.6.6 Halides
- This is a group of minerals which contain halogens, principally chlorine and fluorine. The group includes halite (NaCl) and fluorite (CaF_2).
- Halides are characterized by ionic bonding and are mainly cubic in form, soft, and generally light in weight. They frequently occur as precipitates resulting from the evaporation of saline waters.

Halite, (Rock salt, common salt)
Composition: (NaCl)
Physical Properties
- Crystal systems: Cubic
- Form/Habit: Crystals form cubes, rarely octahedral.
- Colour: Colourless or white if pure, often yellowish, bluish, purplish or reddish tints.
- Cleavage: Perfect cubic {100}
- Fracture: Conchoidal
- Lustre: Vitreous or greasy.
- Transparency: Transparent to translucent to subtranslucent
- Tenacity: Brittle
- Taste: Saline
- Solubility: Soluble in water
- Hardness: 2.0 – 2.5
- S.G: 2.2

Occurrence
- Deposits of rock salt occur as extensive geological beds and are the result of the evaporation of enclosed or partly enclosed bodies of sea water.
- Salt is present in the waters of the ocean and vast inland salt water lakes exist such as the Dead Sea.

NB: Beds of salt differ from other rocks in their reaction to pressure: salt flows while other rocks fracture or fold when subjected to crustal movement.

Flourite, (Flourspar, Blue John)
Composition: CaF_2
Physical Properties
- Crystal systems: Cubic
- Form/Habit: Crystals form cubes, rarely octahedral.
- Colour: Variable including colourless, white, green, purple, yellow or blue.
- Streak: White.
- Cleavage: Octahedral {111} perfect.
- Fracture: Conchoidal to uneven
- Lustre: Vitreous.
- Transparency: Transparent to translucent.
- Tenacity: Brittle
- Hardness: 4.0 (a mineral on Mohs' scale)
- S.G: 3.0 – 3.25

Occurrence
- Fluorite is a late-stage crystallizing mineral in acid igneous rocks. It crystallises at low temperatures in pegmatites and alkaline igneous rocks.
- Fluorite forms the cementing material in some sandstones.

Uses
- The finest grade of fluorite is used in enamelling iron for baths, in the manufacture of opaque and opalescent glasses.
- It is used for the production of hydrofluoric acid.
- The inferior grades are used as flux in steel making and for foundry work.
- Transparent fluorite is increasingly being used in the construction of camera lenses.

2.7 Hands-on activities for Mineralogy

The following activities are centred on the common physical properties of minerals such as relative hardness, taste, solubility and their response to stress.

Activity 1
Aim: To investigate how bond type affects a mineral's solubility.
Requirements: A grain of salt (visible to the

naked eye), a visible transparent grain of sand (or Quartz grain), a coin (or the piece of metal on the eraser end of a pencil), a beaker (or cup), cold water and a timer (a clock or watch).

Procedure:
1. Put the grain of salt, the grain of sand and the coin (or piece of metal) into the cup.
2. Fill the cup halfway with the cold water and allow the set up for 15 minutes.
3. Empty the water from the cup.

Questions;
1. Which of the items in the cup has dissolved (or "dissappeared")?
2. Why did it disappear and why did the others not?
3. Which term in mineralogy will you use to describe the behaviour of the missing item in relation to water?

Activity 2
Aim: To investigate how bond type affects a mineral's response to stress.

Requirements: A visible grain of salt, a visible grain of sand, the piece of metal from the eraser end of a pencil (or a short piece of copper wire), a hammer, a hard surface and a pair of safety goggles.

Procedure:
1. Put on the safety goggles (to protect your eyes from any stray particles)
2. Place the salt crystal, sand grain and piece of metal from the eraser end of a pencil (or copper wire) in turns on the hard surface and hammer them.

Questions
1. Which of the items easily crumbles (breaks into tiny pieces or powder)?
2. Which term in mineralogy is used to describe that physical property?

3. Which of them flattens instead of crumbling?
4. Which term in mineralogy is used to describe that physical property?

Activity 3
Aim: Understanding relative hardness.

Requirements: A black chalk board and piece of chalk.

Procedure: Use the piece of chalk and write the words "**Relative Hardness**" on the black chalk board.

Questions
1. Using the words "relatively harder than" or "relatively softer than", explain why you are able to produce white letters on the black chalk board.

Bonus question: How will you describe the way the following materials respond to bending?
A piece of chalk, the cock of a "Bic" pen (or "Schneider" pen), the cock of a "BEIFA" pen, a rubber band.

2.8 Study Questions On Mineralogy
1. Define the following as used in mineralogy;
   i. A mineral
   ii. Isomorphism,
   iii. Polymorphism
   iv. Pseudomorphism.
5. Explain:
   i. The origin of minerals
   ii. The different types of bonding in minerals and state the effects of each type of bond on the physical properties of the mineral.
3. Describe the main physical properties of minerals.
4. Classify minerals into silicates and non-silicates based on chemical composition.
5. Classify the silicate minerals based on atomic structure
6. Explain the crystallography, physical properties and occurrence of the different groups of silicate

minerals.

7. Outline the differences between amphiboles and pyroxenes.

8. Explain the effects of the structure of silicate minerals on their physical properties.

9. Discuss the Feldspar group and Silica group of minerals under the following headings (or with reference to);
   i. Chemical composition
   ii. Crystallography
   iii. Diagnostic physical properties
   iv. Solid solution (isomorphic series)
   v. Occurrence

10. Discuss the role of Feldspar group and Silica group of minerals in the classification of;
    i. Igneous rocks
    ii. Metamorphic rocks

11. Classify the non-silicate minerals based on their ionic composition and the radical present.

BIBLIOGRAPHY

WEBSITES (all accessed between June 10th to June 26th 2016)

http://etc.usf.edu/clipart/galleries/89-geology.

www.webmineral.com.

www.rockhounds.com.

www.ceritageologi.wordpress.com.

www.adnor.thya.blogspot.com.

www.chemed.chem.purdue.edu.

www.etc.usf.edu.

www.vtuphysics.blogspot.com.

www.bwsmigel.info.

www.chemistryworkbookcrawl.com.

www.chemistryworkbookcrawl.com.

www.tulane.edu.

www.encyclopedia2.thefreedictionary.com.

www.ictwiki.iitk.ernet.in.

www.commons.wikimedia.org.

www.allaboutgemstones.com.

www.tankongvtar.hu.

www.geologycafe.com.

www.slideshare.com.

www.respitirio.utad.pt.

www.lookfordiagnosis.com.

www.thisoldearth.net.

www.hdimagegallery.net.

www.engr.usask.ca.

www.wurstwisdom.com.

www.britannica.com.

www.slideplayer.fr. .

www.dave.ucsc.edu/myrtreia.

BOOKS

Allaby, Michael. 2008. *A Dictionary of Earth Sciences*. UK: Oxford University Press.

Gill, Robin. 2010. *Igneous Rocks and Processes*. Malaysia: Wiley-Blackwell

Gribble, Collin D. 2004. *Rutley's Elements of Mineralogy*. 27th Edition. New Delhi: CBS.

Hammond, Christopher. 2009. *The Basics of Crystallography and Diffraction*. New York: Oxford University Press.

Korbel, Peter & Milan Novak. 2001. *The Complete Encyclopedia of Mineralogy*. Grange Books PCL.

Plummer, Charles, David McGeary and Diane Carlson. 2005. *Physical Geology* 10th Edition. New York: McGraw-Hill.

Whitten, D.G.A. & Brooks, J. R. V. 1972. *The Penguin Dictionary of Geology*. London: Penguin Books Ltd.

INDEX

www.ingramcontent.com/pod-product-compliance
Lightning Source LLC
Chambersburg PA
CBHW081824200326
41597CB00023B/4375